Art Lessons That

Teach Children About

Their Natural Environment

Art Lessons That Teach Children

PARKER PUBLISHING COMPANY

About Their
Natural Environment

Ruth L. Peck

WEST NYACK, NEW YORK

Library of Congress Cataloging in Publication Data

Peck, Ruth L
    Art lessons that teach children about their natural
environment.

    1. Art--Study and teaching (Elementary)
2. Human ecology. I. Title.
N350.P37   372.5   73-741
ISBN 0-13-047415-0

The Scope and Purpose of This Book

Would you like to help the children in your class be more aware of the world around them? Of course you would! That is one aim of all education. But how can you do it—along with the myriad of other things you have to teach?

We live in—indeed, are a part of—an expanding world. We may be intimately involved with only a small part of that world, our own immediate environment, yet from time to time through our travel we come in contact with other surroundings—with people and conditions unfamiliar to us. Frequently, through various forms of mass media, we are made conscious of vastly different areas of this universe, of which the earth is only a small part. Our world continues to grow to almost unbelievable proportions.

Yet are we more sensitive to all the greatness about us, or do we take it so much for granted that we hardly notice it at all? It is possible to go from day to day without ever noticing the changes in the sky above us. How much more meaningful life would be if we were aware of the vastness of a clear blue sky that hints of an endless universe, or the gathering clouds that foretell the fury of a storm. When the children in your class walk among the trees in the park—or past a single tree near the edge of a busy city street—do they go by unheeding, or even deface it? How much greater would be their wealth in life if they were aware of the texture of its bark, the variety of its color overhead, the pattern of light and shadow it creates on the ground below. Is a nearby stream a place to pollute with any trash at hand? Living is richer for the individual who listens to the sound it makes, who knows that there is life in it, who can sit beside it and ponder his own self in relationship to all the other life about him.

Hard as we may try, we cannot live our lives alone. We are related to all nature—to all of the universe. The more we see of it and understand our place in it, the more rewarding it will be to us.

It is the purpose of this book to assist teachers in helping children become more aware of their environment and to be enriched by it. This is not a science book and will not present any scientific facts. Instead, it will help children to see those things which they might otherwise look at without seeing; to be more conscious of the great and the small around them; to appreciate the kinship between man and all nature.

This applies to the children living in the city as much as to children in suburban or rural areas, perhaps more so. Clouds can be seen above city skies as well as country skies if one will but look. There may not be acres of trees in the city, but there are some. Make the children aware of them so that they may enjoy the shade of the trees in the park, the texture of the scraggly ones that grow along the curb, and the changing colors of those in the apartment house yard.

All of us realize that we are living in a world in which technological advances continue to open new horizons in the universe. We know we are living in a world filled with wonder and potential that is still unknown. We study the humanities and learn of ecological facts, yet our actions fail to keep pace with our knowledge, and so our lives are dulled and blunted.

Man, however, is a creative being who can enrich his life through his own actions and sensitivities. The elementary teacher, of all persons, sees both the need and the possibilities for creativeness within the child. But how can it be brought out and developed? This book suggests ways in which creative art can help the child to see, to understand, and to appreciate the world about him. Each lesson will help the child to think about, and interpret through art, some part of his environment. In this way his knowledge will be sharpened and made more meaningful by his own creative endeavor.

We know that children learn best when they take some active part in reacting to those things about which they are learning, and so it is hoped that the creative art lessons will be a valuable asset to the student and the teacher. These lessons may be the catalyst that begins the learning process in the area of science or social studies, reading or creative writing, or, they may be the culmination of study that has preceded them. They may be activities that help children see and be aware—and so enriched. Use them as they will best fit your need.

The author of this book is an elementary teacher—a specialist in art education. She has been an elementary classroom teacher as well, and she understands your problems and your needs. She brings to you a book which will help you meet those needs.

Art is the basis of all living, for it helps us to see and then to express our ideas in visual form. It helps us to understand and to communicate that understanding to others. It helps us to find purpose in life by seeing our relationship to all our world.

It is my hope that the practical and easy-to-follow lessons in this book will assist you to help the children in your class see themselves and the world about them more clearly. It is my hope that, through their creative expressions, the children will understand themselves and their relationship to their environment in such a way that new horizons will open for them—horizons which will make their lives richer and more meaningful.

Ruth L. Peck

Other books by the author:

What Can I Do for an Art Lesson?
Art Lessons on a Shoestring
Art and Language Lessons for the Elementary Classroom

Contents

The Scope and Purpose of This Book 7

PART ONE—EARTH

1   People 19

     lesson 1—Only One of Me (self-portraits)   21
     lesson 2—I'll Be Like That (tempera painting)   29

2   Animals 35

     lesson 1—What Would You Be? (printing)   37
     lesson 2—A New Kind! (cut paper)   43

3   Bugs 49

     lesson 1—Little Bitsy Bugs (wire and pipecleaner
             bugs)   51
     lesson 2—Said the Spider . . . (stick and yarn
             constructions)   55

4   Trees and Plants 61

     lesson 1—Real but Unreal (trees)   63
     lesson 2—Up from the Earth (dry brush painting)   69

5   Rocks and Stones 75
     lesson 1—It Used to Be a Stone (living things)   77
     lesson 2—Hard as a Rock (painted rocks)   81

6   City 85

    lesson 1—Roofs Against the Sky (newspaper pictures)   87
    lesson 2—In a Hurry (real and abstract)   93

7   Country and Suburbia 99

    lesson 1—Sh! Listen! (transparencies)   101
    lesson 2—If You Could Plan It (styrofoam
           construction)   107

PART TWO—SKY

8   Flying Things 113

    lesson 1—Up in the Air (collage)   115
    lesson 2—Cloudy Weather (cotton batting clouds)   121

9   Weather 127

    lesson 1—Forecast . . . (cut paper)   129
    lesson 2—Blowing Up a Storm! (chalk drawings)   135

10   Air and Wind 139

    lesson 1—Blow, Blow, Blow (blow painting)   141
    lesson 2—Windy Weather (mural)   147

PART THREE—WATER

11   Animals of the Sea 153

    lesson 1—Down to the Sea (chalk and charcoal)   155
    lesson 2—That's Where They Live (cut paper)   161

12   Plants 167

    lesson 1—Feel the Sea (fingerpainting)   169
    lesson 2—Underwater Garden (stitchery)   177

13   Oceans, Rivers, and Lakes 183

      lesson 1—Drifting Along (cut paper)   185
      lesson 2—At the Marina (box construction)   191

14   Shells and Sand 195

      lesson 1—New Fossils (plaster of Paris)   197
      lesson 2—A Lonely Beach (mixed media pictures)   203

PART FOUR—THE UNIVERSE

15   Earth 211

      lesson 1—We Live Here (collage)   213
      lesson 2—It Can Be Beautiful (litter collage)   219

16   The Sun 225

      lesson 1—Shadow Weather (changing shapes)   227
      lesson 2—Fun in the Sun! (tempera painting)   233

17   The Moon 239

      lesson 1—Moon Play (circle designs)   241
      lesson 2—Trip to the Moon (box painting)   247

18   The Stars 253

      lesson 1—Twinkle, Twinkle Little Star (chalk stencils)   255
      lesson 2—How I Wonder What You Are! (legend
                pictures)   261

19   Other Planets and Space 267

      lesson 1—Bugs Can Be Good (imaginary bugs)   269
      lesson 2—Way-Out People (box space men)   275

Index 279

Art Lessons That

Teach Children About

Their Natural Environment

part one
Earth

1 People

lesson 1
Only One of Me

OBJECTIVES

1. To become more aware of individual differences in appearance of people.
2. To observe the placement of facial features.
3. To experiment with a combination of common materials.
4. To become familiar with the work of Georges Rouault.

Oh, no! You're not like me. There's only one of me!

Do you like to watch people? When you're in a crowd it's fun just to look at people, isn't it! It's fun because they are all different—different in many ways.

We all like to dress differently from other people—it wouldn't be much fun if we all dressed alike all the time, would it? But clothes aren't really a part of us. What is there about us that makes each one of us look different? Some people have different color skin, and height and weight differ with the individual. Discuss these differences with your class. Let one of the shorter children stand next to one of the taller children, or have one child stand next to you. Some people are taller than others, aren't they? Some people are thinner or fatter than other people, too.

Even if you couldn't see how big a person is, you could recognize him by his face. No two people look just alike, do they? Sometimes twins

21

John

look alike, but if you knew them very well, you would probably find something just a little bit different about them so you could tell them apart. Talk about the things that make people's heads and faces different. Yes, some of you have blue eyes and some have brown eyes—even gray or green eyes. They aren't only different colors, but they are shaped differently, too.

Have the class look at different children in the room. You and you have brown eyes, but see how much darker his are. And look—your eyes are much bigger than her eyes—and yours are rounder than his.

Then talk about other features that make people look different: shape of the face and chin, shape of the nose, color of hair, way hair is styled, straight or curly hair, eyebrows, ears. Encourage all the children to make a contribution to the seeing and discussing.

Many artists like to make pictures of people that show how they are different from other people. One person who did that was the French artist, Georges Rouault. When you first see a painting by Rouault, you may think it strange, because he made his paintings different from those of other artists—just as people are different. There is, however, so much to see in a painting by Georges Rouault that you will be busy looking at all the unusual things—and the longer you look at the painting, the better you will like it. That's true—the more you get to know a person, the better you like him! Such will be the case with Rouault's paintings.

Show one or two reproductions of Rouault paintings. Let the children look at them for a few moments, then ask them the first different thing about them that they noticed. It will probably be the heavy black lines that outline each part of the head and face; another characteristic that will be commented upon will be the strong colors that are used.

Explain to the class that when Rouault was very young he was apprenticed to a maker of stained glass windows. Is there something of the quality of stained glass windows about Rouault's paintings? Yes, the colors are similar—rich colors that in places look as though light is shining through them. Do the black lines make you think of the leaded part of stained glass windows, the part that holds the pieces of glass together? The black lines seem to hold the colors together in a Rouault painting.

The children will want to talk about various things they see. Encourage them to notice and comment about details. No, the face isn't completely realistic, but that doesn't keep you from enjoying it, does it? Yes, the bright colors do shade into darker ones. The black lines have rough edges and are thicker in some places than in others.

Many artists make self-portraits—pictures of themselves. Frequently, like Rouault's paintings, they are not completely realistic, yet they tell something about the artist's appearance. Explain to the children that they are going to make self-portraits in Rouault's style. You'll use different materials, though—the black lines will be done with wax crayons and the colored part with chalk.

Remember that all of you are different; you will want to show that in your self-portrait. Think about yourself for a minute; about what you really look like. Do you have a long, thin face or a rounder one? How do you wear your hair? Do your ears show? Let's see how we will begin our pictures.

Have your class gather around you at a table. These are going to be life-sized portraits, so we will need big paper; 18 × 24 inch white paper will be just right. The first thing to do is plan the size of the face. Make a tiny, black crayon line two or three inches down from the top, and another one about two-thirds of the way down the paper—both of them in halfway from the edge of the paper. Lightly sketch an oval that extends from one line to the other; this is the size and shape of a face. Oh, it isn't the shape of your face; you are not like anyone else, so when you begin your self-portrait you will make it the shape of your face. But see how easy it was to get the size and shape I wanted! I'll leave it a light sketch for a few moments while I plan the hair. Does your hair come down over your forehead or is it pushed back? Is it short so that your ears show—or does it come down below your chin and over your shoulders? I'll make the hair on my person like this, but you make yours the way yours really looks. Sketch the line of hair on your picture. Everyone has a neck and shoulders, so I'll put them in, too. The neck is much narrower than the head, like that, and the shoulders round out and down that way. As you talk, run your hands along the sides of your neck to show the width of it, then across your shoulders to call the children's attention to the direction they take. Sketch the lines on the paper.

Now for the eyes and other parts of the face. Where do the eyes belong? Way up there? Let's see. Place one of your hands flat on the top of your head and your other hand under the edge of your chin. Gradually bring them together in the middle of your eyes; your eyes are just about halfway down your head. Move both hands evenly—not just one of them. There they are—halfway between the top and bottom of your head; that is, halfway from the top of the head—not the hairline—to the bottom of the chin. Sketch two eyes, positioned so there's just about the space of one eye between them.

How long is your nose? Well, let's find out. Put the edge of one hand across your eyes and the other just below your chin. Move them together until they meet at the bottom of your nose. Your nose has the same kind of skin over it as the rest of your face—it isn't a separate part like your eyes or mouth. You can just draw the nostrils if you like, or you can draw the whole nose. I'll make mine this way.

Where is your mouth? Try it and see. Move your fingers from beneath your chin and your nose until they come together, a little below the mouth this time. What shape is your mouth? When you draw your self-portrait, make it look like you. My portrait will be like this.

You'll want to show some indication of the kind of clothing you have on. There isn't much space for it, but you can show the neckline of your dress or shirt. Perhaps you have a V-neckline on your dress or a collar or a tie on your shirt. Sketch something simple to finish that part of your picture.

That wasn't hard, was it? The rest of the picture is even easier and more fun. This is going to be done in Rouault's style, so the first thing to do is make heavy black lines around each part of the portrait. Press hard on the black crayon to make one section of your line wide. That begins to look like the leaded line of a stained glass window, but if it is going to look like Rouault's work it will have to be wider and rougher in some places. Don't take the time to do a large area, but make sure there is a decided difference in the width of what you do; make some obviously rough edges to the line. No, they aren't carelessly done—they should be carefully planned to look a certain way.

Are you ready to do the crayon part of your self-portrait? Don't worry about it not looking completely realistic—it isn't supposed to, or it wouldn't be done in the style of Rouault's stained glass type of paintings. At first some of the children may be self-conscious and hesitant to begin. Reassure them over and over again that it isn't supposed to look exactly like them.

The right place to begin is two or three inches from the top. Now, make another line to show you where you want the bottom of your chin to be—about two-thirds of the way down the paper. Before you draw the oval, think what shape face you have—then sketch it lightly. Certainly you can change the shape, if need be. That is why you are using the crayon lightly. That's an excellent beginning! You do have a pointed chin like that, don't you! Don't even look to see what anyone else is doing. This is a self-portrait, so you just want to think of yourself. Oh, we forgot to see where the ears belong, but you can do that for yourself. Put your hands at the top of your ears and bring them across your face. That's right—they meet at your eyes! Makes it handy for those of you who wear glasses, doesn't it! Do the same thing with the bottom of your ears—they meet at the bottom of your nose. Feel the shape of your ear; does it stand out, or does it lay close to your head?

Continue to encourage the children to think of what they look like, but not to worry about their portraits not looking exactly like themselves. When their initial sketch is finished, let them begin to darken, widen, and make rough edges to the lines. You will be surprised at the difference in appearance that results. Occasionally hold up one child's work for the rest of the class to see. Aren't those lines black and heavy! And see how beautifully rough these edges are. Anyone would know whose self-portrait this is, wouldn't he? Yes, it looks like him without being entirely realistic. It is in the style of Rouault!

When you see that each child is well on his way toward a good crayon drawing, give each child a box of colored chalk. It will be better to begin by coloring the face and neck first, then the clothing. This will make it easier to do the outside colors later without smudging any of the picture.

Comment about the color of the skin. Notice that the skin color of Rouault's paintings isn't completely realistic; they look like glass colors rather than skin colors. Some of you have lighter or darker skin than others do and that is all you need to think about. Certainly you can blend colors and, yes, you should even add your freckles. That makes it you!

Now for the exciting part of the picture—the background. Use any colors you like. Use the side of your chalk to create areas of color—some big, others smaller. Make the shapes simple but interesting; don't let them become so fancy that they become more important than the portrait.

Good! I'm glad you're repeating the colors in several parts of the background; that helps your eyes move easily from one place to another. You are doing a wonderful job! Yes, you may blend the colors slightly— and even shade them into darker colors, if you like. Do you think you are getting too many small areas of color? Make some large areas before all your space is used. Better, isn't it?

All at once the portraits will look beautiful. The children will be pleased when you show one to the rest of the class. Now care will need to be taken to perfect some detail that had gone unnoticed before. As soon as a child finishes his work, tack it to the bulletin board where everyone can see it. When everyone has finished and has his work temporarily displayed, encourage the children to make appreciative comments about other children's work. Notice details which identify a child; comment about the heavy black lines that bring out the brightness of the colors next to them; notice the motion and rhythm of color created by its arrangement on the background. Make each child proud of his work. Make him glad to be an individual, different from everyone else.

When you have arranged a gallery of self-portraits, you'll be delighted with it. They'll all be done in the same wonderful style, but they'll all be different. They have to be! There's only one of me—and me—and me, too!

Make It Easy—For Yourself!

1. Primary grade children should become less involved with details of features and techniques, but encourage them to be different—to make it a portrait of themselves, and to make it big. Tempera paint will be a more suitable material than chalk.
2. Children should wear smocks to protect their clothing when they use colored chalk or paint. Push their sleeves above their elbows. A large man's shirt worn backwards furnishes fine protection. Cut off the long sleeves of the shirt so that they don't drag across the child's work.
3. Reproductions of Rouault's paintings are easily available. Most libraries have the large gallery size reproductions. If you cannot obtain any of them, discuss Rouault's work and demonstrate it as a technique.
4. No pencils! Pencils would be a handicap. They encourage children to work small. Do all sketching with black crayon. If it is used lightly, it will not do any harm to the finished portrait, even if changes are made.
5. Chalk may be shared if necessary. If the chalk is new, have the children break each stick in half before they begin work; this does not spoil it, but makes it easier to handle and less likely to break into tiny pieces that have to be thrown away. In addition—it doubles your supply.

lesson 2
I'll Be Like That!

OBJECTIVES

1. To encourage children to be observant of other people about them.
2. To translate a personal idea into visual form.
3. To have experience with an art material that makes possible a rapid and free expression.
4. To gain confidence handling a basic art material.
5. To make the most important part of a picture the largest, and to fill the paper.

Have you ever seen a nurse perform some miracle and said, "I'll be like that!" or watched a high diver plunge gracefully into a pool and said, "I'll be like that!"?

Let's daydream for a while. Pretend you are grown up like your big sisters or brothers, or your mother and father, or like the people you see going to work each day. If you were one of them going to work, what would you like to do?

Some children will have immediate ideas. Several may want to be astronauts. Some will almost certainly want to be teachers or nurses—probably someone will want to be a policeman or fireman. There may even be a skater or a bareback rider in the class.

I am on the stage playing the piano.

Let's think about those things for a while. If you are a policeman you will probably wear a uniform, won't you? Why do you suppose a policeman wears a uniform? Yes, when he is in the street directing traffic it is important that people see him and know why he is there. The uniform tells them that. What are some other reasons why he wears a uniform? Get the children to think of things a policeman does that makes the uniform important, and then talk about details of the uniform. Yes, he probably wears a gun, but it may be out of sight. What are some of the things you would be sure to see? There will be the badge with a number on it, his cap with the same number on it, the shoulder patch with the name of the city on it, brass buttons, perhaps a stripe down the trousers, or there may be stripes or other insignia on his sleeve that tell you his rank. What color is the uniform?

Talk about other people who wear uniforms: firemen, nurses, garage mechanics, garbage collectors, telephone linemen, postmen, baseball players. Why do you suppose they wear uniforms? Try to get the children to see some purpose to the person's work, some detail of his activity.

Who are some people who don't wear uniforms at work? That's right, it isn't necessary for teachers, salesmen, or secretaries to wear uniforms, is it? These people don't look any different when they are working than they do at any other time. How would an artist tell us that a person is a teacher? Well, he might show a class of children in the picture with the teacher, or the teacher might be writing on the chalkboard or helping a child at her desk. An artist would have to show where the teacher was and who else was there. Do you think that would be a good idea in a picture of a policeman, too? Yes, he might be directing traffic, helping at an accident, helping some children cross the street near a school, or riding in a police car. Something would have to be in the picture to show where he was and what he was doing.

Encourage the children to express many ideas. Each time, relate the kind of person to a picture of him. Where would the football player— or artist, acrobat, or clerk—be? Who else would be there? What would he be doing?

Could you put all those things on one piece of paper? Yes, I think you could. They wouldn't *all* have to be small; in fact, one of them should be very large. What should be the largest part of your picture? The biggest thing should be the most important thing. In a picture of a diver, the diver would be the biggest. His feet could be at the top of the paper and the tips of his fingers could be at the opposite side, on the bottom of the paper. Would there be enough space to show the diving board, the water, and perhaps a crowd of people watching? Certainly! There would be plenty of space around him—and all of those things could be small.

Yes, even the water could be small, because you wouldn't have to show all of it; it could be at the bottom edge of the paper.

Take plenty of time to talk about many things that people do when they work. Mention some things that children don't think about. Encourage them to make things as visual as possible by describing details, action, and the surroundings.

Let's really pretend we are these people—someone you would like to be when you grow up. But right now you will have to be an artist, so you can paint a picture of yourself as that farmer you would like to be, or the garage mechanic you want to be, or you as that clown, or that mountain climber, or the doctor—or whatever it is that you want to be. Think of just what you will look like in the picture and what you will be doing —and if anyone else will be with you. But remember you will be the big, important part of the picture. Plan to put yourself in the picture first so there will be plenty of space for you. Yes, you might be so big that only a part of you will show. If you are a teacher writing on the blackboard, it won't be important for your feet to show, will it? Have you ever seen a photograph taken that way? In a good photograph, the important things are up close to you.

Distribute painting supplies to the class and let each child go to work. Walk about the room to help each child in any way he needs.

Now that's a good beginning! That's as big as you could be if all of you is going to be on the paper, but there will be plenty of space all around you to put other things later. Oops—don't you think you had better let that be somebody else who is unimportant and make another you—a big one? Now I can see why you made the car first! It is going to be a police car with you inside. I'm glad you made only the front of the car on the paper; that makes it big enough to see what you look like—and the rest of the car isn't important, is it? You can put the flashing red light on the top and part of the word *Police* on the door; that's enough to tell us what it is. A forest ranger? You will have lots of trees in your picture, too, won't you? You could have a bear in the distance. Are you going to paint something in that empty space? When your picture looks just right, don't add one more thing. An artist has to know when to stop painting.

After all the paintings have been finished and the supplies have been put away, give each child an opportunity to show his picture. Perhaps someone else in the class can tell what the child is going to be. Then let the child who painted the picture tell anything else about it that he wants to. Call attention to the large size of the person who is important. Notice the details that explain who he is and what he is doing. Children love to pretend—and to paint—so each experience should be a rewarding one.

When all the paintings are on display your class will have a problem. One minute they will want to be like that—or maybe they'd rather be like that—or that or that!

Make It Easy—For Yourself!

1. You want the children to have lots of ideas and to be creative, so allow plenty of time for them to think and share ideas. Direct their thinking into many areas. Ask questions and make comments which will help them recall experiences they have had, to help them see more clearly people they perhaps have only casually observed until now. Add something new to their ideas in order to increase their thinking.
2. Organize the class for painting into groups of from three to five children. Have an empty desk for each group where they can put the materials they will share—paints and water—and where they can leave their brushes when they are not using them.
3. Cover all work or supply areas with newspaper.
4. Have a small can half full of water on each sharing desk so the children can wash their brushes in it.
5. Egg cartons make ideal paint palettes. Half a carton will provide space for six different colors. In this way even the youngest children can carry all the paint for their groups at one time, and the carton is disposable, so no washing of paint containers is necessary at the end of the lesson. Children can provide an ample supply of cartons.
6. Plastic squeeze containers make ideal paint dispensers if your paint does not come packaged that way. Paint left in them remains moist and ready for use from one lesson to another.
7. Assign a helper from each group for that lesson. The helper will (1) cover all desks in his group with newspaper, (2) give each child a piece of 18 × 24 inch newsprint (easel paper), (3) put enough brushes on the sharing desk so that there is one for each child, (4) get half a can of water for the sharing desk, and (5) put a carton of paint on the sharing desk.
8. No pencils! No preliminary drawing. Think and then paint.
9. All children should stand to paint. They will have greater freedom of motion when they stand, so they will make better pictures with fewer accidents.
10. Upper grade children will prefer to use large watercolor wash brushes if they are available, in place of easel brushes.

11. Clean up the easy way. Have the same helpers collect the brushes from the sharing desks. Leave them on a newspaper at the sink so they can be washed later. Have the helpers bring the cartons of paint to the sink, empty the surplus paint into the sink (there will be little waste), and stack the cartons on a newspaper. Wrap them in the newspaper before discarding them, so no paint will spill into the wastebasket. Helpers bring the cans of water to be emptied, rinsed, and returned to their storage area. Helpers should fold the newspaper on the sharing desks and put it in the wastebasket. Then each child will stand to pull his newspaper out from under his painting, fold the newspaper, and place it on the sharing desk. Helpers then put the piles of newspaper in the wastebasket.

12. Younger children should bring the paint and water to you at the sink, but older children should take care of all the work themselves.

13. Store the brushes flat in a box or standing on their wooden ends when they are not in use. Never leave them standing on the bristles.

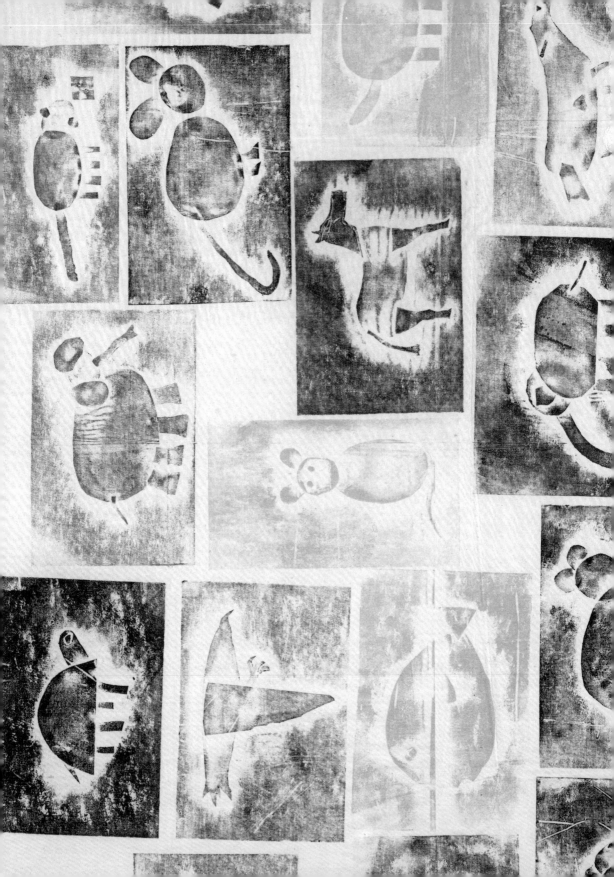

2 Animals

lesson 1
What Would You Be?

OBJECTIVES

1. To be more aware of the differences between animals.
2. To introduce a printing technique.
3. To provide an opportunity for a cooperative project.

If you weren't you, what would you be?

Suppose there is a magician who could change you into an animal—any kind of animal. He tells you that you could have your choice of what animal you would become. Oh, sometime he'd change you back into you again, but for a while you would be an animal. What kind would you like to be?

Yes, your pet cat has a very pleasant life, doesn't he? You give him plenty to eat—and maybe he even catches a mouse or two on his own. Yes, I'm sure a pet dog has just as nice a life as a pet cat. Would you want to be a tiny dog or would you rather be a big watch dog? A seeing eye dog is a hard working dog, isn't he—but he gets good care, too.

Would all of you want to be pets of some kind, or would some of you rather be wild animals? Really? Why would you want to be a giraffe? Talk about a variety of animals that children have read about or seen in a zoo or circus. Some children will change their minds rapidly and decide they'd like to be first one animal and then another. Yes, if you were a

37

Mouse

mouse I think some people would be afraid of you, but you'd have to be careful not to be caught in a trap or by a cat before you could be changed back into you again!

Some of you may want to be many kinds of animals, but the magician will only change you into one animal, so decide which kind you would most like to be. There's just one thing about it—you are going to be the magician, so you will have to know just what that tiger looks like —or that raccoon, or that St. Bernard, or whatever it is that you are going to become. You wouldn't want to become a lion that no one could recognize; if you were a mouse and no one knew it, they wouldn't be afraid of you!

Talk for a while about the things that are distinctive about each of the animals. Does a lion's tail look like the tail of an elephant—or a poodle —or a Siamese cat—or a monkey? What makes a camel look different from any other animal? What does his neck look like? His head? His tail? His legs? Encourage each child to take part in the discussion.

Let's see how we're going to do this. We'll really use magic and print a picture of you as an animal. First we'll have to cut a picture of the animal out of thin cardboard; oaktag will be fine for that. As you talk, begin to cut the thing you are going to make. Hm-m—no one is going to be a turtle, so I'll make that. The shell is flat along the bottom and then rounds over the top. A turtle shell is low, so I'll just make a low curve, but I have to remember to make it big enough to look right on this size cardboard.

Yes, my turtle is going to be glued to the heavy cardboard. When you do this kind of printing, you make a raised picture and attach it to the background. When my turtle is glued to the heavy cardboard I will have a *printing plate*—that is what you call the thing you use to print. We will see how to do that later, but right now let's finish this turtle.

As you continue, talk about the small head that pokes upward from the front of the shell—and the tiny tail that sticks out from the back of the shell. Cut out each part and arrange it on the cardboard. How could I show that the shell has a rough, square-type design on the top of it? It won't do any good to draw it on the shell; that wouldn't make a raised edge, and it wouldn't show on the print. I could cut a few square shapes, or a few lines and attach them to the top. Or I could cut some lines into the carboard. See—when they are cut into the oaktag they leave the shell as a raised surface. Are there long legs on a turtle? No, no, no! They are so tiny and short that a turtle can seem to glide along without showing his legs at all. Cut two tiny legs—the other two wouldn't show on this picture —and paste the parts to the cardboard background.

Now let's see how we print this. Cover a work area with newspaper. Squeeze a small amount of waterbase blockprinting ink on a piece of glass and use a *brayer* (roller with a handle) to roll and spread the ink in all directions until there is an even layer of ink. Then roll the inked brayer over your cardboard plate. Be sure to roll the brayer off all the edges of the plate so they, too, are inked, but don't roll the plate any longer than necessary. Will I try to ink every bit of the printing plate? Well, not really. I'll roll the inked brayer over all of it, but the parts of it that are close to a raised surface won't get any ink on them. That's fine; if the whole plate were inked, I'd just get a solid color print—there wouldn't be any turtle on it at all. Let's see what this will look like when it is printed.

Lay your printing plate on a piece of white or light colored cloth. Place a clean paper (slightly larger than the plate) over it. Rub it hard with the side of your fist. Rub around the edges of the plate first to make sure the print has a sharp, clear outline around it and pull the plate from the cloth.

Look! There's the turtle on the cloth. Notice the definite edge around the print—that is an important characteristic of a print. Always be sure to rub your fist around the edges of the plate when you print it.

Can you see why we don't try to get ink on every speck of the cardboard plate? That's right! It would have made the print a solid shape instead of the shape of the raised cardboard—a turtle in this case. There wasn't any ink up close to those raised edges, so you can see the shape of each piece—even the tiny lines of the shell show.

Someone may notice that the picture on the print faces in the opposite direction from the one on the plate. That's because the plate and the cloth were facing each other when they were printed, so when they are taken apart they are facing in opposite directions. Demonstrate it with your hands. Place both palms together—the thumbs are together, but look —when I open my hands my thumbs are pointing in opposite directions.

Explain that more prints could be made by re-inking the plate and printing it again and again. You will print on cloth each time, but not always on small pieces like this. The second print will be made on a big piece of cloth that is large enough for each of your prints. It will be like a mural, and everybody will print one part of the mural. The room will be a busy place while the children prepare and then print their plates—you will be kept busy, too.

Good! That really looks like a lion! The mane around his head and that kind of tail wouldn't let him be anything else, would they! Would it be better to cut the eyes out of the larger piece than to add such tiny specks? Little pieces are apt to pull away from the plate when you print it. Don't glue anything to the plate until you are sure everything is just the way you want it, and be sure all the edges are glued tightly to the background—you don't want your plate to come apart when you print it. Choose a white or colored piece of cloth as soon as you are ready to print.

As you walked about the room you could have given each child a tube of glue and a sheet of scrap paper. Also, put the supply of small pieces of cloth where the children can choose the color they would like. Prepare three or four inking areas in one or a number of parts of the room. By limiting the number of inking areas you will be able to better control the activity there.

When a plate is inked, the child should return to his own desk to print it. Have him choose his cloth first, and have it all ready at his desk before he inks his plate; this will keep the cloth clean, and will allow him to print while the ink on the plate is still wet and in good printing condition. Place the plate in the center of the cloth so that later the surplus material can be cut away, leaving only a narrow border of cloth around the print.

After a child has made one print on a small piece of cloth, have him re-ink his plate and bring it to the area where you have the large piece of cloth on which all the children will print. Have the first child place his plate close to, but not touching, one corner of the cloth. A supply of clean paper should be within reach so that he can place it on top of his plate to rub it (otherwise bits of ink would be rubbed onto the cloth, and prints would be smudged as others were rubbed close to them). A narrow space should be left between prints so that none of them touch.

Look carefully at each child's first print. Was it inked all the way to the edges? Was it rubbed thoroughly—and to the edges—so that all of it printed? Don't try to correct what may not be a satisfactory print, but be aware of the problem and what should be done to improve the next print.

Find a safe place where each print can be left to dry. Tack the mural to a bulletin board so that all the work can be seen at one time. Comment about the variety of animals that are printed, the differences among the same animals, the details that are especially good to describe certain animals, and good printing techniques. Make it a successful experience by showing appreciation and approval; let children add comments about their own or other children's work.

If you weren't going to be you but were going to be changed into an animal, which one would you want to be? There are lots of choices!

Make It Easy—For Yourself!

1. Have your class gathered around while you demonstrate.
2. No pencils! Cut the shapes without any preliminary drawing.

3. You may want to have all pieces of heavy cardboard for the background the same size so they will fit together more easily.

4. If printing is a new experience for your class, talk about things that are printed: newspapers, magazines, books, cards, posters, and maps.

5. Cover all work areas with newspaper. It won't do any harm if the ones at the inking areas become covered with ink, but the ones at the children's desks should be clean to prevent glue or ink from getting on the print. Let children get clean papers if they are needed.

6. Use any kind of cardboard—preferably heavy—for the background. Thin cardboard such as oaktag is best to make the parts of the picture; it is easy to cut and makes a sufficiently raised edge. Tiny pieces tend to pull away when they are printed. It is better to cut away small areas from the larger ones than to add to them.

7. Let the glue dry before inking the plate.

8. Use waterbase blockprinting ink. It is easy to wash from the brayers and glass with clear water. You may want to use only one or two colors of ink. It is the printing experience that is important, not the choice of color.

9. Pieces of glass about 9″ × 12″ make good inking plates. Cover all edges of the glass with masking or adhesive tape so no one will be cut.

10. The large piece of cloth—white or light-colored—should be large enough to allow each child to make one print on it. If there are a few empty spaces left to be filled, let the class choose which ones should be repeated to fill all the space.

11. Children should wear something to protect their clothing from the ink. A man's large shirt with the sleeves cut off and worn backwards makes a fine smock.

12. Do all inking at the inking areas and all printing—except on the mural—at the children's desks. This will enable the work to proceed faster, with children having to wait their turns less often. Supervise the mural area carefully, to be sure no child spoils the work of the rest of the class.

13. Remember—the plate is made of cardboard, so with use the parts tend to lift and come off. Roll the ink over the plate only enough to cover it. Continued rubbing tends to destroy the plate and make printing difficult.

14. You may prefer two lessons for the project—one to discuss the lesson and make the plate, and a second lesson to do the printing.

Nike. He wiggles and eats grass and doesn't like to fight.

lesson 2
A New Kind!

OBJECTIVES

1. To stimulate children to think imaginatively.
2. To express an idea in visual form.
3. To encourage children to make their art work distinctly their own.

Sooner or later everybody gets tired of the same old thing. Be different—make a new kind!

You've seen all kinds of animals, haven't you? What are some you have seen?

There will be lots of answers. Ask leading questions to remind children of other animals. What are some other pets you have seen? What animals have you seen in a zoo or circus? What are some wild animals you have seen in the woods—or even in your backyard? What is the smallest animal you have ever seen? What is the biggest animal you have seen?

That must be almost every animal there is! No, there are several more kinds—the ones you are going to make. They are going to be different from any other animal that ever lived. They are going to look different and have a different name—but most of all they are going to have a personality that makes them do something that is different. They will be animals, but they will have personalities just like people do.

What are some personality characteristics that people have? What is a characteristic that gives a person a pleasing personality? Certainly, you like a person who is friendly. What are some others? Kind, cheerful, helpful, honest, happy, humorous, gay, loving, friendly, generous. What kind of personality wouldn't you like? Right—just the opposite of these. What personality characteristics might a person have that would make you afraid of him? You would be afraid of a person who was vicious, or cheated, or wasn't dependable, or who punched or kicked other people. He would be a good person to stay away from, wouldn't he!

You can't see these characteristics, but you know a person has a good personality by the way he acts. He acts happy; he does generous things. He might show his personality by smiling or laughing—by looking pleasant. Can you sometimes tell by looking at a person if he is a good-natured person, or if he is an angry or vicious person? Sometimes his personality does show, doesn't it!

Can you imagine a strange new animal who had a personality? Yes, sometimes your pets have personalities, but they are hard to see just by looking at them. The animals you are going to make are kinds that no one has ever seen. You are going to create the first one of its kind—he will be an animal that does something differently than any other animal that has ever lived, and he will have a personality that shows.

There will be questions, so take time to answer them. How big will they be? Why, any size you want them to be. All real animals aren't the same size and these don't have to be, either. Oh, it can't be as small as a mouse, or it might get lost; it certainly can't be as big as a lion. Show the class the size of the paper that is available and explain that if they want their animals larger than this, they can paste several pieces together. Just don't make it too big to handle or too small to see easily. No, of course it doesn't have to be the color of a real animal—these aren't real animals. Perhaps you can use colors that will help to show their personalities.

Think about your new animal for a minute. Decide what kind of a personality you want him to have. What kind of thing would he do? Is it something good or something bad? Would you like to have him around, or would you want to stay a long way from him? What shape head and body would he need to be able to do that? No, don't tell me about any of these things. Keep them a secret so that your animal will be different from anyone else's. Would he need a tail? What kind? Would he need any special kind of legs? Or any particular number of them? What would they have to look like?

Continue to ask questions which will direct the children's thinking. Take time to answer children's questions as they arise. Do you know what your animal is going to do? What personality he has? What he is going to look like? Fine! Let's begin to make them come to life.

Give each child a pair of scissors while groups of children take one or two beginning papers. This will get everyone started quickly with the materials he needs to begin.

Decide how big your animal will be—neither too tiny nor too huge. Don't let him look like any animal you have ever seen or heard about; this animal has to do something special that shows his personality. It looks better to put contrasting colors on top of each other; they wouldn't show if similar colors were put on top of one another. He looks pretty cocky, doesn't he! Is that an important part of his personality? Good—it looks that way! It may look better to use many colors; that depends on what you're trying to show. What an ugly personality! The colors and shapes both say so. I wonder what he does to make him that way; you have to be ready to tell us something about his actions, too.

Have each child clear away all his supplies as soon as his new animal is finished. Then let him write on the back of his animal a few notes about it—his name, the kind of thing he does, things about his personality. It will help him to organize his thinking so that he will be able to tell the rest of the class about it.

Let groups of children take turns showing their animals to the rest of the class and telling about them. Later you will want to plan an exhibit of all the work. You may want to display them with such titles as: Nice to Have Around—or Stay Away!

So you thought you had seen every kind of animal! Here are some new kinds that never even existed before!

Make It Easy—For Yourself!

1. Have a wide assortment of either 9″ × 12″ or 12″ × 18″ colored construction paper.
2. Urge children to take only one or two beginning colors—the kind they will need more of than any other color. Encourage them to share smaller pieces of other colors that are left over. If they need another large sheet of paper or a color that is not available from another child, they should return to the supply area for it.
3. When you see that the children have a good beginning, give each one a paste brush, a bit of paste on a scrap paper, and a piece of newspaper to do his pasting on.
4. If paste brushes are not available, each child can make a paste

applicator by folding a scrap of paper several times until it is a narrow strip about a half-inch wide. Bend it in the middle to give it added strength.

5. As soon as several of the children begin to paste their parts together (after all the animal has been arranged), walk about the room to collect unwanted scraps. Carry a 12″ × 18″ paper and let children pile flat pieces of scraps on it. Empty the pile into the wastebasket occasionally; this will give the children more usable work space, as well as make the final cleanup easier and quicker.

6. Don't paste the animals to a background. Leave them as separate animals.

7. Plan a creative writing lesson when the children can verbally express something about their animal. Let each child express his ideas in any form he chooses, or plan to teach some new form of expression. It is good to combine visual and verbal expression.

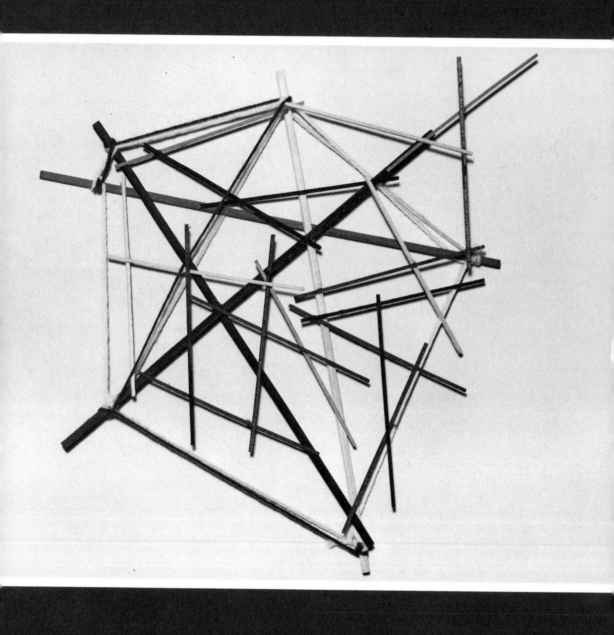

Seven-Legged What-Do-You-Call-It

3 Bugs

lesson 1
Little Bitsy Bugs

OBJECTIVES

1. To combine realism and imagination.
2. To experiment with a combination of flexible materials.
3. To create a balanced three-dimensional picture.

How big is a bug? Have you ever seen giant bugs—or little bitsy bugs? I wonder why it is so many people don't like bugs.

There will be protests from some children that they *do* like bugs, but others will gesture their disapproval. Why don't you like them? All bugs aren't harmful; some of them destroy the ones that do damage. They would be good bugs, wouldn't they? Have you ever looked closely at a bug? Some of them are very pretty. Did you know that bugs have six legs? No wonder they can move so fast! Have you noticed the different parts of their bodies?

Encourage children to describe bugs they have seen. Some older children may have considerable information about them; give them an opportunity to tell some interesting facts to the rest of the class.

If you were going to make bugs out of wire or pipecleaners, what could you do with it to change it from just a long line? Yes, you could bend it into any shape you wanted. Is there any way you could make the wire or the pipecleaners stronger, or even more solid? Certainly, you could. Two pieces twisted together and then shaped would be a lot stronger than just one piece. You could bend it so tightly together that it makes a solid mass; you could bend the line around your finger or a pencil. Then, when it is removed, it stays in a coil—and you can bend that coil in any direction you like. As you talk, bend and shape the wire.

Talk about bugs they can make. They can look like real bugs or they can be some special new bugs that you create. Yes, if they're real bugs they will have six legs, but they might be so tiny they would hardly show. As children think about their work, they will have questions; answer them, and ask other questions of your own which will further their thinking and help them clarify their ideas. Real bugs come in all sizes so yours can, too. Yours will have to be bigger than real bugs, though; even one pipecleaner is bigger than the largest bugs—and perhaps you will use several of them, or wire. You may want to use some of both; that would give your bug a more interesting texture.

Coated doorbell wire comes in a variety of colors, as do pipecleaners; have several colors of each material. Let each child choose one or two wires to begin with. Have a pair of wire cutters available in case it is necessary to cut the wire.

It looks like you are going to make all three sections of its body a different color. That is a good idea and will remind us that bugs do have three parts to their bodies. It is hard to make two sections attached so strongly that they don't fall apart; perhaps if you overlapped the wire and the pipecleaner a little more and then twisted them tightly together they would be stronger. If you want yours to grow an extra leg and have seven of them that will be all right—it just won't be any bug you have ever seen before, so make it entirely different in other ways, too. Good! You have made the wings look solid, and very important to your bug. Well, they have to be if they are really going to fly, don't they! Oh, there's the second pair of wings that some bugs have; they wouldn't help it to fly, would they!

Continue to comment about the children's work and to help them in any way they need. Some children will know exactly what they want to do and finish it without any problems; others will need to be encouraged and will require more time to experiment. Urge them to do their own thinking and to make their work different from that of other children.

Do you have a name for your kind of bug? No, bugs don't have the kinds of names that you do—or that your pet has, but every kind of bug has a name. Yes, they are long names sometimes. If you have a different kind of bug, make up a different kind of name for it.

When all the bugs are completed, have six or eight children display their work together at an area you have cleared. Arrange them carefully, so that each bug shows well; don't put him too close to someone else's bug. Then let the rest of the class gather and look at them; talk about the variety, the things that are different, the way the materials have been used. Comment about the ones that look like some common bug. What is there about it that tells you what kind it is? Which ones are imaginary bugs? What tells us that it isn't real?

Then let another group of children display their bugs—and have the class talk about them. A piece of 12″ × 18″ white or colored construction paper makes a good display background for each group of bugs. The constructions are small, so not a great deal of space is required to display all of them.

You never liked bugs before? Well, you'll like these—all of them, from the giant ones to the little bitsy ones.

Make It Easy—For Yourself!

1. Precut some of the wire into pieces about eighteen inches long; this will enable children to choose their beginning materials quickly. Let them return to the supply area for more wire or pipecleaners as they need them.
2. Wire (pipecleaners are wire, too) should never be cut with scissors. Beside being difficult or impossible to do, it nicks the blades and ruins the scissors for their correct use. Teach children the proper use of and respect for materials. Wire should be cut with wire cutters—most pliers have a wire cutting edge. Two pairs of them will be all you'll need—leave one pair with the supplies and keep one with you. As you walk about the room, cut off the wire or pipecleaners where children indicate they want them cut.
3. Encourage children to find other information about bugs. Some children may want to look up information about a specific kind of bug and then write a report about it. Let them read their reports to the class or have them available where other children may read them. If children bring in books about bugs from the library, make them feel they've made a valuable contribution. Show your appreciation for the extra interest and work. Do some children have collections of bugs—or would you like to begin a class collection of them? Art can be the starting point of continued learning.

A Pretty Trap

lesson 2
Said the Spider...

OBJECTIVES

1. To encourage children to be more observant and to find beauty in the common things about them.
2. To experiment with line to create space and shape.
3. To combine materials with different characteristics to create a similar effect.

Said the spider to the fly—stay out of my web!

There is something that spiders do that no other insects do—they spin webs. Your mother doesn't like it when a spider gets in the house and makes a web on a lamp in the living room or across the kitchen window; she gets rid of it quickly, doesn't she! Usually those are just little webs, maybe just a single strand—but have you ever noticed the big spider webs?

Children will have stories to tell about spider webs in old, abandoned houses or in barns; about the times they ran into them and were covered with silky threads that clung to them; about giant webs that get bigger as each child tells about the ones he has seen.

What shape is a spider web? Are you sure they're always round? No, they can be any shape; it depends upon where the spider spins them. If they're in the corner, they are the shape of the corner; if they're out in the open, they may be round—or almost so.

Have you ever looked at a spider web that the sun was shining through? Lovely, isn't it! Each little line seems to be a tiny piece of silk that shines in the light. No, your mother may not want them in the house, but they can be lovely.

Did you ever wonder how spiders are able to build such delicate things, yet make them so strong? They seem to start at one point and grow out and out from there. Is the smallest part of the spider web right in the middle? No, sometimes it is off to one side—just the way you sometimes put the most interesting part of your picture slightly to one side to make it more important. When you make a picture, do you make everything exactly the same size? No. Have you noticed that the parts of a spider web are different sizes, too? Some spiders seem to be more clever than others and make spider webs in more interesting designs than those made by other spiders. Yes, they must be better artists!

Talk about the lines that reach out—and the lines that connect and hold them together. Some lines of the spider web are straight while others seem to curve; some of them are thicker than others. The lines make different kinds of shapes inside them.

We're going to make spider webs. The lines won't be as thin as those of a real spider web, but they will be even lovelier. Show the class the supplies they will use—colored sticks of various lengths, colored yarn, and glue to hold the sticks together.

Have your class gather around you in an area covered with newspaper to protect it from the glue. Take three or four of the longer sticks (about a foot long) and cross them in various ways. It wouldn't look particularly pleasing to have them cross in the middle, would it? Move them about until you have created areas of different sizes and shapes. There's the beginning of a spider web, but I'll have to put a drop of glue on each point where they attach to another stick. No, the ends don't have to connect—just place them so that they look right.

Would you like to help to make this spider web? Let first one child and then another select a stick and find a good place for it on the design. A different length stick makes a pleasing new line, and see—it has changed the shapes that were there before. Yes, certainly you can make it larger, but be sure the spaces inside the design are also made interesting. Glue takes a while to dry, and the sticks will come apart easily while the glue is still wet, so leave the spider web flat while you are working on it. Well, let's not take time to finish this one, but can you see how you can make your spider web grow bigger and prettier at the same time?

Let each child take ten or twelve sticks, a newspaper, and a tube of glue. Remember, spider webs take the same shape as the place they are made, but yours can be any shape you would like it to be. It may look

partly square or almost round, or it may have a quite irregular shape.
Don't try to make all of the outside shape before you make the inside
part; build the whole spider web at one time. Certainly, you may use
many colors or you may use only one or two. You decide what you want
your design to look like. Good! That looks square without all the sides
being exactly alike. Looks better with some variety, doesn't it! Oops!
Don't try to pick it up—it would surely fall apart! When it is completely
dry it will be very strong.

We'll let these dry for a few minutes while we put the glue and
extra sticks away. Leave the newspaper where it is so as not to disturb
the spider webs, and have your class gather around you again.

You are smarter than spiders, so let's do something they can't do. So
far the spider webs are made entirely of sticks, but we're going to add
one more material to them. As you talk cut off a piece of yarn two or
three feet long. Yarn has a different texture from wood, so it will add
variety of texture to the design. If texture is a new word to your class,
explain what it means. Even if your eyes were closed you could tell by
touching it that it is a different material. The feel of it—the texture—is
different.

But back to this design. Will you use glue to hold the yarn to the
spider web? Perhaps it could be done that way, but yarn is soft and
flexible, so we can tie it. Tie one end of the yarn to one of the sticks. See—
I could make the line go over there to that stick—and see the new shapes
it creates. I could make it go to that stick—and it has made the spider web
bigger by adding another space on the outside. I could wrap it around
those sticks where they cross and then bring it back out to that stick. Now
it has changed the shape of the yarn line, hasn't it. Or it could go to that
stick—or that one—or that one—or to any one at all. Which way do you
think would look best? All right—let's tie it right there. The sticks are almost dry so they are fairly strong, but you will still have to be careful
with your design.

Make two or three more lines with the yarn. Let different children
decide each time just where the yarn should go. Yes, it does look better
than it did before. Yours will look better, too, but do you think it would
be possible to make too many yarn lines? Yes, it would—so look very
carefully, and when you think it is just right don't add another line.

Let each child select one or two pieces of yarn and then return to
his own desk. Can you find a place that looks a bit empty? That would
be a good beginning place for the yarn. Tie it to one of the sticks and
then move it to various places to see where it looks best. Oh, that was a
fine place for it! It divides that big space into two others, making it look
much better! Don't handle your design too much; it is still not completely

dry and it would be dreadful if it came apart. It's a good idea to have all the sticks of one color and all the yarn a contrasting color.

Continue to compliment and encourage the children until all the spider webs are finished. Have a section of bulletin board cleared and covered with white paper. One after another, as they are finished, tack them to the bulletin board. Place a tack through two or three pieces of yarn at different parts of the design and into the bulletin board to hold it in place. Don't crowd them—leave plenty of space around each one so that it will show off to its best advantage. Have each child's name on a tiny, separate piece of paper, so that it can be placed with the spider web when it is displayed. Comment to the rest of the class as each one is put in the exhibit. Comment about the arrangement of lines, the overall shape, the shape and variety of size of the open areas, the placement of the yarn lines, the good workmanship, the three-dimensional part of a few of them, anything which is different.

No spider webs were ever as pretty as these, were they! Real spiders may like to trap flies in their webs, but can you imagine wanting a fly in one of these!

Make It Easy—For Yourself!

1. Colored sticks come in various lengths, from a foot long to only an inch or two long. Some are very thin, others about a quarter of an inch square. Have as much variety as possible.
2. If colored sticks are not available, use regular round applicator sticks and toothpicks. Yarn will add color to them. Display on a black background to afford maximum color contrast.
3. Precut several pieces of yarn of each color. When children need more, let them return to the supply area and cut off whatever they need.

4   Trees and Plants

Tall and Lovely

lesson 1
Real but Unreal

OBJECTIVES

1. To encourage children to be more observant of detail.
2. To demonstrate that a picture may be both realistic and abstract.
3. To develop ability to sketch with rapid, rhythmic motions.

Someone tells you a thing is real—but unreal. Which do you believe? Why, both, of course!

You've seen thousands of trees, haven't you? You must know what they look like—but do they all look alike? Well, they all have a trunk, and branches, and leaves in the summer, but they all look different, too. They're like people—all people look alike in some ways, but you have to look carefully to recognize one particular person, because each one is different, too.

What are some of the ways in which trees differ from one another? Someone may start by telling you the names of some kinds of trees, but they are just the names we call them. What do they look like that makes them different from other trees? If you are in an area where you can see several trees from your classroom windows, ask the class to look at them. Are they all alike or can you see different kinds of trees? How do you know they are different? What is there about them that looks different?

Start by seeing the most obvious differences: height, and thickness of the trunk. Then notice that trees are shaped differently—that the big branches of some grow upward in long, sweeping, almost vertical lines; that the big branches of others soon swing outward in almost horizontal lines. Do you know of one kind of tree that has branches that droop downward? Right! A weeping willow tree. Comment about trees that are common in your area—elm, live oak, maple, mimosa, poplar, apple, cherry.

All trees, however, are alike in one way. They grow from a tiny seed, so they grow always up—then out—from that. Go to the chalkboard and draw a horizontal line for the ground. We can't see what is going on under there, but we know there has to be a seed and that roots start to grow from it. Then a tiny piece grows out of the ground and a tree has started. If you have ever had a new tree in your yard and have watched it grow, you know that each spring it grows a little taller and tiny branches begin to grow—upward and outward, the exact shape depending upon the kind of tree. As you talk, make your tree picture grow.

Do you think the roots grow, too? Yes, certainly they do. Some trees, as they get bigger, have roots that grow on the surface of the ground where you can see them. They have to reach deeper into the ground and spread out wider as the tree grows so they will get water and food enough for the tree to grow. The trunk is always getting wider and taller and the branches are getting longer, with more smaller branches coming from them.

Let children make comments that supplement what you say and that show they are thinking visually. Each time you add to the tree, start at the ground and thicken the trunk slightly as you move the chalk upward and outward into a new main branch—or along one you have already drawn, ending with a new smaller branch. Trees never grow only on both sides with a hole in the middle; they grow a tiny bit wider at a time, so the trunk is always solid.

As you work, comment about the sketching lines you are making. Do you notice how lightly I sketch—and that the chalk moves rapidly from one place to another? Each line is just one motion. See how the chalk is lifted gradually near the end of the motion. This makes the chalk end with a fine line—the way a tiny branch ends. If this were a particular tree instead of just any tree, I would have to make the branches grow in a special way to make it an apple tree, or a weeping willow tree, or an oak tree, or whatever kind of tree I was making.

Let's look outside again. All those trees don't look alike, do they? Not at all! Can you find the tallest one? Which one has branches that reach out wider than any of the others? Do you see one that has a broken branch still hanging from it? Find the one that has most of the branches growing from just one side. Let the children identify which ones you describe. Then let several children comment about some feature of different trees so that other children can find them.

Explain that they are going to draw one or more of the trees. They will make them so realistic that everyone else will be able to tell which one they have made. Of course you won't be able to put in every branch, but you will look very carefully and draw the main parts of the tree just the way you see them—just the way they have grown. You will draw only what you see—you won't imagine anything this time. It has to be there if you draw it; it has to be real.

There is only one thing that will be unreal—one thing where you can use your imagination—and that will be the color. You can make your trees any color you like. In fact, if you make several trees, each one can be made with a different color chalk, so part of your picture will be real and part of it will be abstract. Artists frequently use this technique.

Give each child a piece of 12" × 18" white drawing paper and a box of colored chalk. Remind them again to look carefully. Pick out one tree you would like to sketch, notice all the details, and then make that tree grow. Later you may want to choose a second or a third tree and make them grow, too.

Make your first tree grow tall enough to fit the paper. If it is a rather short tree, it may be better to use your paper horizontally, so that the tree can spread out sideways instead of upward; it will also leave space on the paper to add other trees. I can tell already which tree you are drawing by that extra low branch. Remember, a tree grows from the roots, so it can never grow down. Certainly a branch may slant downward —but only after it has grown up from the roots. A tree has to get food and water from its roots in order to grow. Good! That's a fine second color to use for the next tree. It is not at all realistic, but it makes a good contrast with the first color. If you are making the tree I think you are, you haven't noticed something—isn't there one branch that makes a rounder line than any of the others? Good! Now you see it. Could you hold your chalk more loosely, so you could sketch more freely? Would you like to make a chalk line for the ground? Make that an abstract color, too. You can see the ground in the distance in back of the trees, so draw the line to make it appear to be in back of them—as though you could walk around them. When you have finished your sketch, go back over each line. Press hard on the chalk to make the color stronger and more interesting.

Continue to help individuals in any way necessary. Occasionally hold up a sketch for the other children to admire—it will compliment the child whose work you show, and it will encourage the others. Help each child to be successful. A compliment, a suggestion, a question, a comment will keep each child looking, thinking, and doing his best work.

When all the pictures are finished, let each child show his work to the rest of the class. They will be delighted when other children recognize trees they have drawn. Comment about the realism of the sketches, the pleasing abstract colors, interesting arrangement on the papers, and good sketching technique.

The display you make of all the children's work will prove it—a thing can be real and unreal at the same time. And lovely, too!

Make It Easy—For Yourself!

1. You may want to take your class outside to do the looking and sketching, or you may plan a trip to a nearby park. If your class works outdoors, see that each drawing paper is taped to heavy cardboard to use as a drawing board. Pressing on the chalk to sharpen the colors after the sketch is finished can be done in the classroom where their desks will make a firmer surface.
2. It is best to teach this lesson in the spring or fall when the leaves are off the trees. If you live in an area where the leaves do not fall, use the foliage as details to be observed and drawn. Use an abstract color for all of the tree.
3. Cover work areas with newspaper to protect them from the chalk.
4. No pencils! No preliminary sketching. Do all the work with chalk.
5. If this is the first time new chalk is used, have the children break each stick in half before using it. Half a piece of chalk is easier to handle than a whole piece—and it is less likely to break into tiny pieces which have to be discarded. In addition, it doubles your supply of chalk.
6. The final picture will be livelier and more pleasing if the child goes over his sketch and presses hard with the chalk to make each line heavier and sharper.
7. Spray a light coat of fixative over the pictures if they are to be displayed where children's hands or clothing could damage them. Hair spray makes a fine substitute for fixative.

New Fern

Dry Brush Painting

lesson 2
Up from the Earth

OBJECTIVES

1. To be more aware of the ordinary plant life about us.
2. To experiment with a new painting technique.
3. To develop ability to paint with rhythmic motions.

Stop! You may be stepping on something precious. Lots of wonderful things grow up from the earth!

Have you ever looked at a thing without really seeing it? Let children tell you several stories of things they looked at but couldn't see—a school book someone looked all over the house for, and there it was on the kitchen table; a baseball bat he couldn't find, but was right where he had looked three times; lunch money he thought he had lost, while all the time it was lying on his desk in plain sight! Sometimes we need extra eyes, don't we!

Let's pretend we have those extra eyes so we can see some things we have looked at many times. You won't have to go far to see them, either. They are all around you—the little things that grow up from the earth. What are some of them? No, not trees. Trees grow to be much bigger than the things we're going to paint this time. What are some small things that grow?

Someone will probably suggest flowers; that's a good beginning. Some of them grow rather tall, but not nearly as tall as trees; some of them are so tiny you don't see them unless you really look for them. Have you ever gone hunting for wildflowers and had a hard time finding them? They're nice to look at, but it's a good idea not to pick them. What are some other small things that grow up from the earth? It won't take long for other things to be suggested: weeds, grasses, ferns, wheat, rye, oats, bamboo, vegetables.

Have the class gather around you. Explain that they are going to use dry brush painting to make the things grow on paper. Do you think you will use much paint when you do dry brush painting? No—and of course, it can't be completely dry either, or it wouldn't paint at all, would it?

Pour a puddle of paint onto a heavy piece of paper. Dip a large watercolor wash brush into the light color tempera paint. Lay the brush flat so there will be lots of paint on the brush, and then wipe the brush onto the edge of the paper so that the surplus paint is wiped off. If there's still too much paint on the brush, wipe it on the paper some more to get rid of the excess paint.

Small plants don't generally grow straight up, so painting them that way would make them look unreal. They bend and move with every breeze—or anything that touches them. I'll just swing a line upward with my brush. As you talk, touch the brush to the dark colored paper, near the bottom, and swing it upward and to one side with one rapid and light motion. Look at the line it painted—is it a solid line, the way it usually is when you paint? No, much of the paper shows through the paint. You can see why I painted on a dark paper—so the dark paper would look good through the dry brush lines. I dipped the brush way into the paint —even though I had to get rid of much of it—in order to get a little paint on all parts of the brush. A little paint on the tip of the brush wouldn't have looked this way when I painted, would it?

Notice that I used the brush lightly—hardly touching the paper but moving smoothly and rapidly. As you talk, make several brush strokes. Move them upward in the same general direction, but make some longer and some shorter than the first one. Bend some, and make some straighter than others. Continue to call the children's attention to the motion and rhythm of the lines as you paint them. Comment about the quality of the lines—fine streaks of paint with the paper showing through them.

Talk about what this could be. It is pretty this way—it looks like a field of tall grass, doesn't it! Could anything be added to part of it to change it into something else? It could be changed into a field of grain or of wild grass that has gone to seed just by adding tiny specks to the top parts of some of the lines. Add tiny diagonal lines to the top of a long line as you talk. They're still motion lines, aren't they? It is like sketching with paint.

What else could be done to change these into other things? Little half rounding lines could be added near the bottom of the paper; then there would be tiny wildflowers in the field. Flowers—single ones—could be added to the ends of these. Suppose there had only been one or two of these lines—what could you have done then? They could still have been any of these things, or they could have become bigger plants and flowers. You could make them into ferns; how would you do that?

Ask each child to think of something small that grows up from the ground that he would like to make. It will fill the whole paper, even though it is a very small thing when it is really growing. How can that be? Well, you could imagine that you have it under a huge magnifying glass, or you can pretend you are a tiny bug looking up at it. Then it would look big to you, wouldn't it!

Give each child a small puddle of paint on a piece of heavy paper. A piece of dark 12″ × 18″ construction paper and a large watercolor wash brush are the only other materials he will need.

Think of what you would like to make before you put any paint on your brush. Are you going to make something tall and narrow enough to enable you to use your paper vertically, or are you going to make a long, horizontal picture? You decide. It's a good idea to try some of the motions first without any paint. You need more paint on your brush than that; such a little bit of paint on the end of your brush would make a thin line, but it wouldn't look like dry brush painting. That's better! Now wipe off most of it. That was a good smooth motion; I can tell by the appearance of the line that it made. Hold your brush loosely and the line will be loose and free, too. Good! One big, lovely flower is fine! Are you going to add a stem of the same color, or would you like another color for that? Don't put too many things in your picture. You can make a painting very quickly when you do this type of painting.

Take a minute or two to look at several of the paintings. Talk about the freedom of motion shown in one, the simplicity yet completeness of another, the variety of ideas that are shown by others.

Would you like to make another dry brush painting of some kind of plant life? Decide what you would like to make this time. Take a good look at your first picture; see the things that are especially good about it so that you will be able to do just as well next time. Are there some things you can do better? Were you holding your brush a little too tightly so that the lines look stiff and tight, too? Did you try to draw shapes with the paint instead of just motions? Well, next time lift the brush off the paper after each motion. Did you have too much paint—or not enough—on your brush? Did you put too many lines in your picture? You can fix any of those on the next painting.

Give each child more paint and another piece of paper. Walk about the room as you did before, reassuring and complimenting as each child requires. Ask and answer questions as the need arises.

When the second paintings have been finished, let each child fold his palette paper once and drop it in the wastebasket. Brushes can be left on a newspaper at the sink to be washed later.

Have each child put his two paintings on his desk so he can see both of them at the same time. Which one do you like better? Perhaps you will want to hold one and then the other away from you, enabling you to get a better view of them. Which one will you show to the rest of the class? Yes, both of yours are excellent, but you decide which one you think is better. You are right! That one is much freer than the other one.

Let groups of children take turns showing their pictures to the rest of the class. Give them an opportunity to make any remarks about the paintings, and encourage other members of the class to make comments about parts that have been done particularly well. Make it a pleasant time for enjoying other children's work and for appreciating their successes.

You may not have seen some of these things before—but perhaps that's because you don't always see everything you look at. There's no doubt about it—lots of wonderful things have grown up from the ground!

Make It Easy—For Yourself!

1. If you are working with primary grade children, don't stress the dry brush technique. Let them use tempera paint and easel brushes in the regular way, but show them how to move their hands in rhythmic motions.

2. A choice of three or four colors of paint is enough. Use light, bright colors; have black or other dark colored paper to paint on to provide a sharp contrast in color with the paint.

3. It is not necessary to cover the desks with newspaper since so little paint is used that there should be no danger of it getting on the desks. The newspaper would only be a nuisance and tend to pull the palette and pictures off the desks.

4. Old construction paper or oaktag make fine palettes. Pieces about 6″ × 9″ are large enough for the paint and small enough so they don't take too much room on the desks. If a different combination of colors is to be used for the second paper, you may want to change the palettes. Collect the old ones and discard them. Give each child clean palette paper.

5. Brushes need not be washed as children go from one color to another. So little paint will be on the brushes that they can be wiped on a clean part of the palette and put into the second color.

6. Children should stand to paint since standing permits greater freedom of motion.

7. Older children should be able to get more paint for themselves as they need it—only a bit at a time, though. It will be simpler for you to give it to younger children.

5   Rocks and Stones

Squirrel, Frog, and Rabbit

lesson 1
It Used to Be a Stone

OBJECTIVES

1. To develop ability to observe more closely.
2. To encourage children to use found objects as art materials.
3. To experiment with a variety of materials to create a desired effect.

It used to be a stone—but not any more!

Have you ever seen a bug, a bird, or an animal, that looked so much like the things around it that you had a hard time finding it? That is called protective coloration. Sometimes they have shapes similar to the things about them, too. Let children tell about various things they have seen that are examples of this. We don't have any of those things in here, but we do have some stones that want you to think they're bugs or animals, or some other living thing.

Have your class gather around you. Lay a stone—perhaps a flat, oval one—in one hand. Turn it over, then sideways, showing it to the class so they see it from all directions. Does it remind you of anything? Someone may think of a turtle. It does resemble the shape of a turtle, doesn't it! If it were a real turtle it would have a head sticking out one end and a tiny tail at the other end. Would there be long legs? No, a turtle lays close to the ground; sometimes his legs are pulled inside his shell.

Show the class some assorted materials you have: felt, tiny beads, glue, and perhaps a few tiny feathers. Could you make a head, tail, and feet from the felt? No, we're not going to use paint this time—just add bits of things to it, but not cover it. This time we want it to look like a rock, but a rock that is a turtle. Could you think of any way of adding to it to make it look like a shell without covering the whole thing? You could add lines of felt for the designs on the shell.

Place several other rocks on the table. Can you see any different things in these? Someone may see a mouse in one; oh, yes—he'd need a long, thin tail. You could cut a narrow piece of felt for that, and two tiny beads would be fine for his eyes, since mice have beady little eyes. He should have some whiskers, but felt would be too thick for whiskers, and it would be too limp; the bristles of an old brush would be just fine.

Discuss the possibilities of two or three more stones. Lay all of your supply of stones on the table and look them over to see which one you would like to use. Find one which you can make into a bug or animal, or some other kind of living thing. Would you like the stone that looked like a mouse? You don't have to use it; you may choose another one, and someone else may like the mouse.

Give each child who has seen something in a stone the first opportunity to use it; then let groups of children take turns selecting the rocks they would like to use. Put the assorted materials on a desk or table where children may choose from them as soon as they have their stones. Give each child a pair of scissors and a tube of glue.

No wonder you chose that one for a frog! It looks like one even before you add anything to it, doesn't it? Yes, we have some cotton batting you may use for the rabbit's tail. That will make him a real cottontail, won't it! Um-m. That's a very long tail. I wonder what it is going to be. Oh, now I see! It isn't really a tail at all; it is the body of a long snake. I'm glad he is only a felt and stone one. How are you going to make the squirrel's tail stand up against his back and curve over again? It needs to look bushy, too, doesn't it? Well, try it and see if it works. Would you like to have me hold that for you while you add the other piece? Now I see why you were so busy cutting all those points out of felt. A dragon has to have points on his back, doesn't he!

Continue to comment about the children's work and to help them in any way they need. A compliment or a word of encouragement may be all that is necessary.

Gradually, each creature will take on his own particular characteristics. When all of them are completed, have a quick showing. Have each child place his work on a large table and have the class gather around it.

Notice the variety of things shown—there will hardly be two of any one kind of creature. Talk about a stone which was an especially good choice for a particular thing. Notice another with some detail which is just right for that object. Compliment the child who devised some clever way of doing a thing—or who did something unusual—or who solved some special problem—or who had an idea that was completely different from anyone else's.

They used to be stones—or at least you thought they were. But look —they turned out to be turtles and bunnies, squirrels and spiders, and almost anything you can imagine.

Make It Easy—For Yourself!

1. Give each child a small piece of newspaper to work on to protect his desk from the glue. A piece about 9″ × 12″ will be better than covering the whole desk; the larger paper is a nuisance, and unnecessary.

2. Scissors used for cutting paper soon become too dull to cut cloth. It is a good idea to have a second set of scissors that are saved for cloth only. If they are the same size as the other scissors, mark them in some way to identify them. A piece of tape or a bit of yarn tied to the handle is satisfactory.

3. Some children may want to use materials that you have not supplied but which are available in the room—things such as wire, construction paper, or cotton batting. Encourage this and let the children get what they need.

4. Paste will not hold materials to stones; as soon as it dries it will pull apart. Be sure to use glue.

5. No pencils! No preliminary sketching of parts to be cut from felt or other materials. Think and then cut.

6. Display each child's work. A closed display case where the things won't be handled would be fine.

Shapes with New Colors

lesson 2
Hard as a Rock

OBJECTIVES

1. To encourage children to use found objects as art materials.
2. To be more aware of shape and form in natural objects.
3. To create a useful object.

It's lovely! The colors are gay and beautiful. But pick it up—it's hard as a rock!

Let's take a close look at those rocks you have. Some of them are lovely just as they are. Some of them are wonderfully heavy to make fine paperweights or doorstops, but they aren't very pretty right now. Well, that's easy enough to change!

Ask several children to put their rocks on a large table and let all the class gather around them. Can you find a rock that is pretty already? It has interesting colors in it—what could you do to make those colors even more important? If you painted over them they wouldn't show any more. What else could you do? You could paint some areas around the colors that you wanted to show, or just some lines or shapes around them to make them show even more. You might even want to varnish the stone to make it shiny and not change anything else.

Can you find a rock that has an interesting shape but doesn't have any pretty colors? Look at all the different shapes to it! It looks as though it was broken apart from some bigger rock, doesn't it. But what an uninteresting color! What could you do to make it look better? You could paint it. Would you paint it all the same color? Yes, you could. Then it would make the rock's shape more important because the color would look better. You could use several colors and make each space a different color from the shape next to it; that would make all the broken edges important.

Can you find a rock that is smooth enough to paint a picture on it? When you rub your fingers over that rock it is almost as smooth as a piece of paper. What could you paint on it? You could make a duck; perhaps you could make the duck look as though he fits the shape of the rock. You could make his tail feathers round to fit that part of the rock, and the bottom of his body could round there, and then his head and bill could round to fit right there. What else could you fit on it? Talk about many things: a person, a mouse, an elephant—anything at all. Anything you could paint on paper you could paint here; it could be just lines and shapes.

Talk about the possibilities suggested by several other rocks. The rocks are all different, so how you finish them will be different, too. Maybe you will need to do very little—maybe just varnish one—to make it look nice, but you might have to paint all of another rock to improve it. Look at your own rock to see what will be the best way of finishing it to be an attractive paperweight or doorstop, if it is big enough for that.

Each child will need very little painting space, so it will be better to let them work around large tables or groups of desks that have been pushed together to create large work areas. Arrange the painting areas so that the children can share the tempera paints—and later the varnish.

As soon as a child has decided how he wants to paint his rock, let him go to a painting area. Supervise the painting areas as carefully as you would if each were at his own desk. You will be able to do more pointing out of individual work to the rest of the group than you usually can.

Yes, I should think that would be a fine way to paint it. Find an empty space at the painting area and go right to work. Look! There's the rock we talked about. You have used a good combination of light and dark colors to make a fine contrast for each area. Good! I'm glad you've let part of the rock show on yours. Doesn't the light rock color make a fine background for the abstract design on it! Different sizes of polka dots make a lively design; they look like bubbles. Let one color dry before you paint another one next to it, so they won't run together. You don't want the shapes to run together, but you could plan to have the colors run together. You might even wet the rock first to help them blend nicely.

When a child finishes painting his rock, let him leave it at the painting area to dry. Tempera paint dries quickly, but you may prefer to have the varnishing done another time. In any case, when the stones are dry and it is time to varnish them, let children take turns at the work area. One light coat of varnish is enough.

They'll have to be left for several hours before the varnish is dry, but you don't have to touch them to look. Let your class gather around them. Which ones have a picture or color arrangement that looks especially good? Which plans are particularly appropriate to the shapes of the rocks? Are there some that are painted especially well? Which one do you wish was yours?

They're prettier than ever—and still hard as a rock! They'll make wonderful paperweights and doorstops.

Make It Easy—For Yourself!

1. Have each child bring in a stone for himself. Talk about what they are going to be used for, so that an appropriate size will be brought in, or take the class for a walk and have them find suitable rocks.
2. Contribute a few extra stones of your own. If a child decides his rock will be best if only varnished, let him paint a second one from one of your extras.
3. No pencils! No preliminary sketching. Think and then paint.
4. Let children take turns varnishing their rocks. It will take only a minute to varnish each one; fewer brushes will be needed in the varnish. Inch-wide enamel brushes are best for varnishing.
5. Varnish brushes should be washed in a jar of turpentine, then washed with soap and cold water to soften the bristles.
6. Several children can share one supply of paint—and a can of water to wash paint brushes—but each child needs one paint brush.
7. Do not paint or varnish the bottom of the rocks.
8. If you wish, felt may be glued to the bottom of the rocks after the varnish has dried.

6 City

OPEN HOUSE
WOODBRIDGE
Shirlee Elston, Assoc.
MOBILE HOMES 134
MILFORD
GEORGE J. SMITH & SON, INC.
FLORIDA $12,650
FLORIDA UNLIMITED, INC.
617-426-7244
NO SELLING
INCOME OF $300 WEEK UP
HOME INSULATION & ALUMINUM FABRICATING
ARCO
Proftable Opportunity
THE BEAZLEY CO.
Lumber Co.
Morris Seigel Agency 934-8331
HOUSES FOR RENT 114
HOUSES FOR SALE 130
New Haven
F. C. DEVITA CO.
NEW HAVEN NEWLY LISTED
Gianelli Agency 865-2250 or 878-6781
LOW 20s
LYNN WEED REALTY
Joseph Croog Co. REALTORS
Hamden
MADISON
GOULD AGENCY
HAMDEN GRACIOUS LIVING
Orange Sylvan Valley
3 FAM. BEAUTY $31,5
Raymond and Paige Co.
Stevenson Ranch $40s
4 Fam. $29,500
Rosen & Son 389-9574
WOODBRIDGE
THE BEAZLEY
ORANGE
RENTS WAITING
$26,900
PRESTIGE AREA
Real Estate Wanted 148

lesson 1
Roofs Against the Sky

OBJECTIVES

1. To be more aware of buildings crowded together in cities.
2. To experiment with newspaper as an art material.
3. To increase ability to arrange shape and space.
4. To be aware of the importance of background space.

Buildings, skyscrapers, towers, steeples, smokestacks—roofs against the sky! Cities!

Cities are exciting places, aren't they! They're crowded with people hurrying in different directions. They're jammed with traffic—with cars, subways, busses, trains, and airplanes. People shout, horns blow, sirens scream; everything is making noise. But let's get away from the city a bit. Pretend you're in a car just approaching the city, or you're on a boat in the harbor. You're far enough away so that you can't see little things like cars and people. You're far enough away so that you can't hear a bit of noise. You still know there's a city there, though. How can you tell?

Right! You can see the buildings. How do you know it is a city and not just buildings any place? They're tall buildings, and there are lots of them crowded together. They are not all the same. How are they different—even from a long way off? Some of them are much taller than

Overpopulated

others. Do they look as though they are close together in one line? No, no—not at all! They are in back of and in front of one another. Parts of some buildings are hidden in back of other taller ones.

Are the colors of the buildings important? No—the colors tend to look pretty much the same—but there is something different about their shape, particularly the shape of their roofs—the part you see against the sky. Talk about the variety of shapes that are seen against the skyline: the tall, flat-roofed office and apartment buildings; the smokestacks that probe upward from an industrial area; the lower, pointed roofs of smaller buildings; the occasional steeple or dome of a church; the towers, or the unusual shape of some modern building that makes it stand out from all the rest. The buildings cut interesting shapes into the sky, don't they?

Let's make our own cities with interesting skylines. Explain that they will use newspaper to make buildings—tall buildings, short buildings, all kinds of buildings, and lots of them. They will be arranged together— in front of each other, in back of each other, close together, so that they make a crowded city with an interesting skyline. Large 18″ × 24″ black paper will be needed for that.

On the chalkboard, draw a large rectangle to represent the background paper. On it draw a tall building with a flat roof; low on the picture draw a small building with a pointed roof. Let it partly overlap the taller building. You wouldn't see both buildings where they overlap, would you? Which one would you like to have in front? Fine. Then I'll erase the line of the other building. When you arrange your newspaper buildings you will have to decide which one to put in front—on top of the other, that is. Draw several more buildings of different heights and different shape roofs. See—doesn't it make an interesting pattern for the top of the picture?

Which is more important to see—the tops of the buildings or the bottoms of them? The tops are the parts that are different, so they should be the important part of the picture. Point out some parts of your sketch where the sky comes almost to the bottom of the picture. It makes a more interesting arrangement than all solid areas, doesn't it?

Can you make the whole city in one picture? Certainly not! You will be able to see only one little part of the city in your picture, so you will have to decide what part of it you want to show. Does an industrial part of the city look the same as a residential or commercial part of the city? No, there are different kinds of buildings in each area. Talk for a few minutes about what the buildings would look like in various parts of the city. Could there be a park on the edge of a residential area? Yes, you might even show a tree or two to tell us that; you decide.

Give each child a page of newspaper and a black crayon. Sketch some big buildings first, while your paper is large—then make some smaller ones from parts that are left. You may even need a second piece of newspaper. The lines of the newspaper will help you sketch buildings that are straight; you can use them as guides without drawing on them. You want your buildings to have different widths as well as different heights. Don't worry if your crayon lines aren't completely straight; this will be one time you won't mind if they show a bit, so when you cut out the building you can straighten it by going off the crayon line if necessary. Continue to help where you are needed.

Of course, you can do it! An easy way to get started is to draw a completely plain rectangle, tall and rather narrow. See! A fine office building. Now try something a little different—different size, shape, or different kind of roof. Oh, you must be making an old factory area where there are lots of smokestacks. Cut them out right away and begin to arrange them on your black paper. Don't paste anything until you have everything placed just where you want it to stay. Could you separate some buildings just a bit, so the sky comes down low at one point? Good! That makes the other buidings look even taller.

Occasionally show the class a building a child has cut or the beginning of an especially good arrangement. Show the first one or two that are entirely pasted to the background, then stop the class for a few moments.

Do city buildings look like solid walls? A few of them may, but most of them have windows. They show from a distance, too, don't they—but you don't see each separate piece of glass, do you? Perhaps you could use your black crayon to indicate the window arrangement on a few of your buildings. Again sketch on the chalkboard an arrangement of three or four buildings. Sketch parts of the windows of a building making them appear to be in a row, one above the other, all the way up a tall building. Don't make any of them complete—just suggest the size, shape, and spacing. Indicate that the whole front of the building is covered with them. On another building make wide window areas, as though the whole of a floor of the building is one wide window. Suggest the arrangement of windows, but don't draw the whole thing—you are a long distance from the city, you know—you are looking at it from a distance. You may want to use your black crayon to show more distinctly where one building comes in front of another one.

Again walk about the room to help individual children as they need it. Some children may be too timid—encourage them to use their crayons more freely and boldly. Other children may be putting in too many details—urge them to use more restraint. Encourage children to occasionally step away from their pictures, in order to see them at a distance, or take the picture across the room so the child gets a better view of it. Does it look finished or is there something missing?

Do you have a title for your picture? Every picture is improved by a good title. Think of something you want your picture to say—something you have tried to say with paper and crayon. Then put it into words —in a title—on the back of the picture.

Have a showing of each child's work. Comment about the kind of area that is depicted, the appropriateness of the title, the arrangement of shapes and the arrangement of black paper spaces that are left, the good workmanship, and the effectiveness of the crayon lines. Find something which tells each child that he has learned and been successful in at least one area. Later, display all the pictures.

It's fun to visit a city and be a part of the noise and excitement, but it's good to get away, too. It is nice to see it quietly from a distance. Skyscrapers, steeples, towers, smokestacks—all a part of the city.

Make It Easy—For Yourself!

1. Use the parts of the newspaper which contain fine print—classified ads, financial section. Avoid large print headlines, advertising, and pictures. You are trying to create a textured effect and so need an evenness to the page.
2. Teach children to paste on all the edges of their work. This keeps the work flat and prevents curling or loose edges that would indicate poor workmanship. Do not paste in the middle of an object or in spots.
3. If paste brushes are not available, satisfactory paste applicators can be made by folding scraps of paper until they are narrow strips, about an inch wide. Bend them in the middle to strengthen them.
4. As you walk about the room while the children cut out their buildings, let them give you their leftover pieces. A page of newspaper will make a fine tray for them to lay flat pieces on. Once in a while, empty them into the wastebasket. They will have lots of scraps, and this will help keep them cleared away.
5. The big 18″ × 24″ black paper and the newspaper require a big work area. Extra space can be given each child by having him stand to work. In this way, he can keep his arrangement on his desk and his other supplies on his chair. Make use of other work space, too—empty desks, tables, counter space, even the floor.
6. Encourage each child to think of a creative and meaningful title. The title should be put on the back of the picture—never on the

front where it would detract from the picture. Use the title on a small paper with the picture when it is displayed.

7. Pictures may be displayed individually, or you may want to touch one against the other to create the effect of a skyline mural.

Confusion

Real and Abstract

lesson 2
In a Hurry

OBJECTIVES

1. To translate the noise and confusion of a city into visual terms.
2. To experiment with line and color to create an emotional effect.
3. To use chalk as a fluid material to create a mood.
4. To demonstrate that a picture may contain both realism and abstraction.

Everybody's in a hurry! You have to be, too!

Have you ever been to a big city? What was the first thing you noticed? There's lots of traffic in a big city! There are automobiles and trucks of all kinds—and what else? See how many kinds of vehicles the class can name that might be in a big city. Taxis, motorcycles, trains, subways, airplanes, busses, ambulances, fire engines, police cars—all making noise. Horns are blowing, brakes screeching, motors roaring, sirens screaming. Noise, noise, noise—another thing you notice about a big city. The bigger the city, the more noise! But those things don't make noise all by themselves; there have to be people. Are the people just in the cars and busses and other things on the street? They are everywhere!

They're on the sidewalks, in the stores and offices, in the parks, crossing the street—pushing and shoving, always in a hurry—so there is confusion and noise. And, yes, there is pollution!

Explain that they are going to make pictures that will create the mood of excitement—of noise and confusion. You'll use chalk for that. What is a noisy color? Certainly, colors can be quiet or noisy—just like people can—but they *look* noisy instead of *sound* noisy. Red is the noisiest color of all! Red shouts and screams just as people and sirens do—it is a good city color. Orange can be almost as noisy as red. Any bright color is noisy—especially bright red and bright orange. Black is a strange color; it makes other colors stronger—and noisier. It can help to say smoke and dirt, too.

Would a city be all red and orange and black? Probably not—even in a noisy city there are some moments that are quieter than others. If it were the same noise all the time, you wouldn't hear the brakes and the sirens, would you? So you will be able to use some yellow or purple or blue or other colors. The contrast will help the noisy colors sound even louder.

How will you put on those noisy colors? What shapes will they be? What is a noisy shape? Some shapes are quiet and some are noisy. Use your hand in the air to make an almost straight line. Is that a noisy line? No, no! That horizontal line hardly made a sound at all. Move your hand in a slow, rolling, rippling motion. Noisy? No. You could only hear that if everything else were quiet—certainly not in a noisy city. Well, what kind of a line or shape is noisy, then? Jagged lines—abrupt changes of direction —points! All of those can be heard above the noise of traffic and people and sirens. They are the noise of traffic and people and sirens! Can you imagine the noise they would make if they were red or orange, or even if they were other colors with some black between them?

You are going to make a whole page—a whole city—filled with noise. Use color, shape, and line to make it as noisy as you can. There won't be anything real in it—just noise. It will be as though you closed your eyes so you couldn't see the people or traffic or buildings. All you would do is hear the noise. When you make a picture, however, you have to hear with your eyes. All ready to be noisy with your eyes?

Give each child a piece of 12″ × 18″ white drawing paper and a box of colored chalk. Begin anywhere—a noise can start anytime, anyplace. Make it a big sound—so a big shape. Good! That's a noisy shape. Now what color will you put next to that orange noise? Good! That will surely make a clashing sound. You have two light colors next to each other. Too quiet, isn't it? A black line between them will sharpen them and make them both noisier. Every speck of the paper has to be filled with color. Noise goes everywhere, so the color—the picture noise—has to go everywhere, too. Can you push a little harder on the chalk to make it heavier?

As each child finishes using the chalk, let him put it back with the supplies. While other children are finishing this part of the picture, begin to talk again about some of the things that make a city noisy or crowded or confusing: buildings, automobiles, taxis, oil trucks, trailer trucks, ambulances, police cars, fire engines, and people. Yes, these are going in your pictures, too—but not with chalk.

When many of the children have finished the chalk part of their pictures, stop the class. Hold up several of the pictures that are particularly noisy, that express the sound of a city. Compliment each child for a good arrangement of color or for the expressive shapes he has created.

This much of the city you could hear and know about even if your eyes were closed, but now pretend you open your eyes. What would you see in all this noise? Certainly, it could be any of those things that might make the noise—the police cars, the automobiles, the trains, the motorcycles. Yes, there would be people, too—lots of people. You would see buildings; you couldn't go to a big, noisy, crowded city without seeing buildings.

If you put all these things in your picture there wouldn't be any noise showing, so you will have to decide which two, three, four or so of them you want to include in your picture. It might be only people or only any one kind of thing, or you might want to include a variety of things. Think about it and decide just what you want to include in your picture.

Show the class the assortment of colored paper which is available. You do not need many colors, but include red, orange, black, and a light color such as yellow. Others may be added if you want them.

There will be questions, so take time to answer them. It isn't necessary to make things the color they really are, just use colors that will look good where you want them to be. You will paste your objects right on top of your noisy paper. You aren't making a realistic picture, so there is no ground or sky. A person might be near the top of the paper instead of at the bottom. Would you put a red car on a red chalk area? No, it wouldn't show there; you would either make it a different color or you would put it in a different place. Don't paste anything to the background until you have arranged all the objects. You can paste the parts of a car together, but don't paste the car to the picture until everything else is in its place.

Continue to help each child as the need arises—with materials, with encouragement, with suggestions. Urge him to be different—to do something which is uniquely his own.

It was fun going to a big city. Everybody was in a hurry! You can still hear the noise and see the confusion!

Make It Easy—For Yourself!

1. Cover all work areas to protect them from the chalk—and later from the paste.
2. No pencils! No preliminary drawing, either for the chalk or cut paper parts of the pictures.
3. Take plenty of time to motivate the class to make sound and confusion into visual form. Interchange sight and sound words.
4. The side of the chalk will cover a large area quickly, but the end of the chalk will make the color more vivid. Paper will absorb only a limited amount of chalk; the rest remains on top as surplus. Once in a while, the children should drop the extra chalk off the paper—or gently blow it away (if they are not too close to other children).
5. Have the colored construction paper cut into 6″ × 9″ pieces. Let each child choose two pieces—this will get everyone started in a hurry. Children may then share leftover pieces with other children, or return to the supply area when necessary.
6. You may want to divide the work into two lessons—one for the abstract chalk part, and a second for the cut paper, realistic objects. In that case, it would be a good idea to spray each chalk picture with a light coat of fixative to prevent undue smudging.
7. If paste brushes are not available, fold scraps of paper several times until they are narrow strips, about a half-inch wide, and bend them in the middle to strengthen them. They make fine disposable paste applicators.

7   Country
and Suburbia

lesson 1
Sh! Listen!

OBJECTIVES

1. To interpret in visual terms the mood of a place.
2. To experiment with a combination of transparent and opaque materials.
3. To use a combination of common materials in a new way.

Sh! Listen! Don't make a sound. You have to be quiet to hear it.

Let's pretend you take a long walk way out into the country. You're all by yourself—there isn't another person in sight. What would be one of the first things you would notice about it? Lower your voice as you talk.

Probably one of the first things said is that it's quiet. That's just right! You would especially notice it if you had just come from the city. The city is a noisy place, but the country is quiet.

If you listen carefully, you will hear some things—things you wouldn't hear at all unless you were very still. You will certainly hear birds singing. What else? There might be a cricket or a katydid—or a rabbit hopping through some tall grass.

Would some sounds you hear depend upon where you are in the country? What might you hear if you were on a farm? You might hear cows—or pigs or sheep. You could hear chickens—or roosters, ducks, or geese—there are all kinds of animals to hear. Would you hear only little sounds on a farm, or would you perhaps hear any loud sounds? You might hear the motor of a tractor or some other farm machinery, but you would hear it as one particular thing—not as all the mixed-up noises you hear in the city.

There are other sounds you would hear on a farm. Perhaps you can think of some different ones, but let's go to another area of the country—to a forest. What could you hear there? There might be the sound of leaves falling to the ground. You would hear the sound of crackling twigs and leaves and pine needles as you walked on them; that is a woodsy sound.

Talk about other sounds they would hear in the woods—a crackling campfire, wind whistling through the trees, a stream bubbling over the rocks, the occasional snap of a small twig as a squirrel jumps from branch to branch. There are other sounds in the forest, too, but you will have to be quiet and listen for them.

You decide where you would like to be in the country. Then be very still and listen for sounds. Which ones do you especially like? Which ones would you like to put in a picture?

These will be such quiet pictures that even the person who looks at them will have to listen to hear their sounds. Explain that they are going to make pictures of the place where they hear a country sound. Somewhere in the picture will be something to hear.

Let your class gather around you at a table. Tear off a piece of waxed paper about two feet long. Fold it so that the ends curl away from each other, and lay it open on the table. We will put our pictures inside the waxed paper—so will we arrange the parts on both halves of the paper— or on only one half? Good! On only one half. Then the other half of the waxed paper can be folded over the top of it, and the picture will be inside.

We'll pretend this is a farm. I might show that it is a farm by putting in a barn or a tall silo. You pretend this piece of black paper is one of those things—whichever one you like. It could even be a part of the farmhouse. Let's put some animals on this farm. What kind would you like to hear? Chickens? Fine. They could be cows, if you would prefer. Well, you pretend these small pieces of paper are animals, or ducks, or whatever you like. Group them together and move them about to change the arrangement. Which way would your chickens look better—or your sheep? They could be slightly overlapped as long as you can still see what they are. They look fine that way, don't they?

You will use only black paper for whatever you cut out to put in your pictures. The shapes will be silhouettes against a transparent background, so color won't be important. It will be important, though, to cut each thing in just one piece. Don't cut the ears of your pig separately or cut the legs of the cow out of separate pieces of paper. Can you tell why each thing should be cut as one whole piece? Well, let's wait until later and you will see the reason for it.

Explain that there is a way of adding color to the picture—to make it more pleasing and to help show that the sounds are in the country. Green is a good country color, so I'll add some to show that these ducks or goats or horses are out in the fields. I'll have to use a transparent color—wax crayons will be just fine for that. Remove enough of the paper wrapper to enable you to use an open pair of scissors to scrape off bits of the crayon. Let it drop onto the waxed paper where you want the ground to be. Brown would have been a good color for the ground, too; even some yellow would be fine to add to the green.

Certainly, let's add some blue to the sky. Should any of the crayon be put on the black paper? No, the color of the crayon wouldn't show on the solid paper. Can you tell why it will show on the waxed paper? Well, maybe we'll have to wait to see about that, too. Just be sure not to put on too much crayon.

Fold the empty half of the waxed paper over the half that has your pretend picture and crayon shavings on it. It would fall apart if it were picked up this way, wouldn't it? Slide a piece of cardboard under the waxed paper and carry it to a previously prepared ironing area. Cover it with a piece of newspaper and iron it with a medium hot iron.

Hold the transparent picture up to the window. Wouldn't that be a nice picture if it were a silo, or a house, or a barn, with chickens, or horses, or whatever you want them to be? And look at the lovely colors! Do you see why we use crayon instead of paper? The crayon melts and spreads, and lets the light go through it. Why was it important to cut each thing as one whole thing and not in parts? When it was ironed, the separate pieces might move just a bit. You wouldn't want the roof of the barn blowing off—or the cow's tail to separate from his body!

Each child will be excited and eager to make his own picture. Remind them once more to think of where in the country they want to be and what they are going to hear. Give each child a pair of scissors and a piece of 9″ × 12″ black construction paper. As you walk about the room to see what help each person needs, give each child a piece of waxed paper.

Fold the waxed paper so the ends curl away from each other—so that the bottom half lays flat on your desk and the top half rolls out of the way. Decide what will be the biggest part of your picture and cut it while you still have a large piece of paper. Good! That's a sound no one else heard. You had to be very quiet to hear that, didn't you? Don't worry about your rooster not looking exactly like a rooster. Give him a pointed beak and some tail feathers—and two thin legs with separate toenails for feet. We will be able to hear him just fine. Maybe he will be standing on a fence and we will know he is crowing. What a wonderful place for a camp! You will be able to hear all kinds of quiet sounds there—especially at night. You could use black crayon to tell us it is night. Shave off bits of crayon on your picture any time you are ready for it. Sunrise! Good for you! That's when roosters make the most noise, isn't it!

As children begin to come to the ironing area, supervise it carefully to make sure the iron is handled safely. Keep fingers out of the way, and make sure the iron is always left standing on end.

Let each child hold his picture against the window, so he and the rest of the class may see it. Encourage him to comment about the sound. The rest of the class—and you—should find something to compliment him for, so he feels a sense of pride in his work. Be genuine in your comments, however, so the children won't think you are insincere.

When all the pictures are displayed on the windows—or are backed with white paper if they are shown elsewhere—you will hear all kinds of pleasant sounds. You will have to listen, though, for they are quiet sounds you would expect to hear only in the country.

Make It Easy—For Yourself!

1. If words such as "transparent" and "silhouette" are unfamiliar, explain their meanings. Demonstrate each one.
2. Cover the desks with newspaper to protect them from the specks of shaved crayons—it is easier than to try to remove the bits of crayon.
3. Use wax crayons only. It is the wax in the crayon which melts and spreads to make transparent colored areas—and helps to hold the picture together.
4. Have the children arrange their pictures so the bottom will be at the fold of the waxed paper. This will prevent anything from falling out of the bottom. It is best not to let any of the construction paper parts touch the sides. Keep a sealed edge of waxed paper on both sides to prevent anything from coming apart.
5. The piece of newspaper over the picture protects the iron from stray bits of wax crayon, and it prevents the crayon from spread-

ing too much. Iron over the picture only once or twice. Too much ironing can blend the colors so much that they become muddy in appearance, and can remove so much of the wax from the paper that it doesn't stick together.

6. A fine ironing area can be made with a pile of opened newspaper thick enough so the heat won't penetrate to the surface beneath it.

7. Three or four pieces of cardboard to be shared are satisfactory. The limited number of pieces control the number of children who can be at the ironing area at any one time.

8. If you have a paper cutter, use it to trim the ragged edge of the finished picture. The ends of the wax paper that were slit by the metal cutter on the box are certain to be slightly rough and uneven.

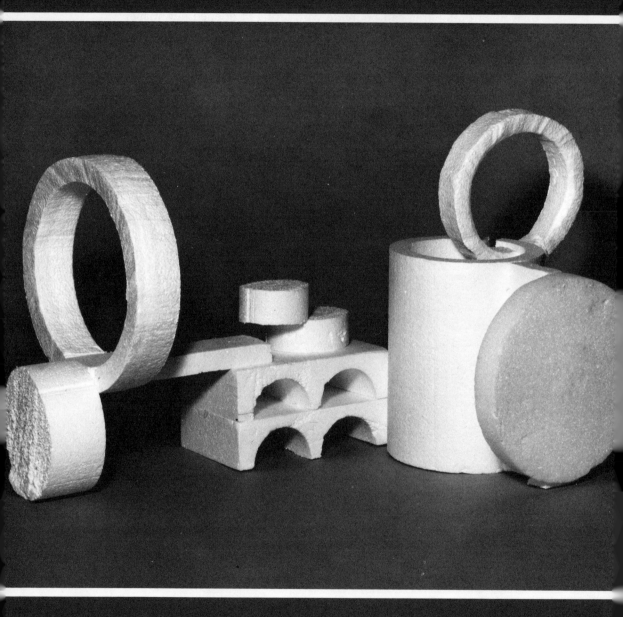

Theater and Museum

lesson 2
If You Could Plan It

OBJECTIVES

1. To interpret modern buildings as space and mass.
2. To provide opportunity for a cooperative project.
3. To experiment with the use of scrap material for an art activity.
4. To have experience working on a large project.

Why did they do it that way? They should have moved this over here—and put that over there. How would you do it if you could plan it?

How would you like to have the mayor tell you that the town is going to have a new civic center, and that you are on the committee which will plan it? It would be exciting, wouldn't it? It might even be a little frightening—it would be a big responsibility.

Probably at some time you've thought of something you wished the city had. If you could arrange for the city to construct, in one new area, all kinds of buildings that you think the city should have, what would you want? Swimming pools and skating rinks may be suggested first; those would be fine things to have. Lots of people would enjoy swimming or skating there. They would be indoors, so you could go swimming even in the winter, and skating in the summer. That would be fun, wouldn't it?

But remember, you are on a committee that is planning this civic center for everyone in the community—and everyone wouldn't want to swim or skate. What are some other kinds of buildings that you would have for them? Talk about the various interests of people of all ages. Gradually, such things will be suggested as a theater, library, sports arena, museum, restaurant, art gallery, hotel. It might be nice to have a building where people could go to take part in many small activities—to sew, to paint, to play ping pong, to hear a lecture—or just to visit with each other. List the suggestions on the chalkboard as they are made.

Well, that's your job—to plan a new and wonderful civic center where all kinds of people can go to do the things they enjoy. It is a big job, isn't it? Too big! Let's do what other committees do when they have a job that is too big for one group to plan. They break up into smaller groups, and each one of them plans just one part of the project. You could do that, couldn't you?

Choose one child and ask him to pick a committee of one or two others. Pick people who will work well together. A committee won't get much done, you know, unless each person helps. So choose someone whose work you respect and who will get along with the other people on the committee. Fine—you should have some good ideas. You get together and talk about which part of the civic center you would like to construct.

Continue to organize small committees until each child in the room has at least one other person to work with him on a project, then ask everyone to stop their planning for a few minutes. The whole committee— all of you, that is—have to decide some important things. The first thing we all have to know about are the building materials that will be used.

Show the class the assortment of pieces of styrofoam you have. Have as much variety in size and shape as possible. Many breakable items now come packaged in chunks of styrofoam that fit the contour of the objects. Inquire in advance among friends, store owners, appliance dealers, and others, and ask them to save the styrofoam packing for you. You will be able to collect big blocks, cylinders, cones, odd shapes of all kinds, solid pieces, and pieces with holes through them.

The variety of sizes and shapes of the styrofoam will be fascinating and help children's imaginations go to work. Isn't that a great piece! Wonder what someone will do with it. Does anyone have an idea what you could do with this one—or this? You will have to decide what you want your arena or skating rink or hotel to look like, and then you will need to find the right building materials for it.

A special glue is needed to hold styrofoam together. Ordinary white glue or quick-drying cement may tend to eat into the styrofoam and erode it. It can be used in an emergency—sparingly—but styrofoam glue will be more satisfactory.

You might want to use an occasional piece of construction paper to show some detail—you could use ordinary pins to hold it to the styrofoam—but don't let the paper, or other minor materials that you might decide to use, become important. We are not working with color this time—just mass and space that it creates. You will want to make each part of your construction an important part of the whole. Look at it from all directions to see that the arrangement of solid and open areas is pleasing.

Go from one group to another to see what they are planning to make. If one group plans to make the same kind of building as another group, suggest that they change to another. You may find it helpful to check off on a chalkboard list each thing a group plans to make. This will show each of the committees if they must make a second choice. There will be questions, too, as you go from one group to another. Take time to answer them, or if it is a basic question, talk to the entire group about it.

Someone wants to know about changing the shape of some of these building materials. Well, yes, you can change them somewhat, but not a great deal. We don't have big tools to change the big pieces, but perhaps you could use the open blade of a pair of scissors to make some minor change. If you have other simple cutting tools, show them—and set up safety and sharing rules for their use.

The room will be a busy place—as any construction site would be where many new buildings are being created at the same time. You will be the general superintendent and will be as busy as anyone could be. There will be problems with materials, labor, management—but no strikes!

That's a wonderful hotel! I'm glad you are making it different from any you have ever seen. A new kind of civic center needs new kinds of buildings, doesn't it? Do you think you are putting those two sections of your museum too close together? Would the space be more interesting if you moved them just a bit? Try it. Um-m, a long, low series of buildings. They must all be related in some way. Good! That's an excellent plan for a hobby center. Each kind of acivity has its own area, but it is easy to go from one to another. Are you having trouble making that balance? That's one thing architects and builders have to consider. Could you put something under that section to support it? Stick a pin through it to help hold it in place until the glue has dried.

All at once things will begin to take shape, and one building after another will be completed as each committee finishes its work. Take time to look at each one, and let the builders tell anything they want to about the function or style of their particular construction. There will be lots of wonderful ideas!

If possible, find one large location where all the buildings may be displayed together as one modern civic center. Label each part and include the names of the committee members.

You did have a chance to plan it—and it's just the way you want it. Someone else may ask why you did it that way, but you know. You like it that way!

Make It Easy—For Yourself!

1. You will need a large amount of styrofoam. Begin collecting it well in advance of the time you want to use it. Cardboard boxes of various sizes and shapes can be substituted for the styrofoam, but they are much less intriguing.
2. A small saw or two would be handy to have in the room to cut through some of the larger pieces of styrofoam. A paring knife or an ordinary table knife makes a good cutting tool for smaller pieces. Young children should make constructions as they would with building blocks—without changing their form.
3. You may want to take two lessons to complete the buildings. Not only would this provide more time, but it would enable the children to find some extra and particular piece of styrofoam that is not available during the first lesson.

part two

Sky

8
Flying Things

lesson 1
Up in the Air

OBJECTIVES

1. To be more aware of the variety of things which move through the air.
2. To experiment with an assortment of textures within a picture.
3. To increase ability to organize and arrange parts into a pleasing whole.

Remember the time you were unsure about things, and you said everything was up in the air? Well, you'll be completely sure of things this time—but everything will still be up in the air!

Do you think of the sky as just empty space? I suppose there are times when you look up into the sky and there doesn't seem to be anything at all there. But if you look about a bit you'll probably see something. Maybe there are just a few clouds; clouds belong up in the sky. Of course, birds belong there, too. You see birds on the ground and in trees, but you wouldn't be surprised to see them in the sky.

So far you have thought of something that the air holds up and moves. Clouds are held up by the air, aren't they? But birds have to move themselves through the air. They are both natural things, though. Can you think of something that you might see up in the air, but is a mechanical thing? Airplanes are machines, and their engines push them through the air. What are some other things of that kind that fly through the air? Certainly—there are rockets, helicopters, dirigibles.

Let's go back to things that the air holds up by itself—things that just float through the air. Kites and balloons will be suggested immediately. Perhaps you will be able to think of some more.

What are some living things that you might expect to see in the air? Ducks—and geese and seagulls—and owls—all kinds of birds, but there are other things, too. Mosquitoes fly about and are a nuisance, aren't they? There are flies and bugs and beetles of all kinds; there are butterflies, too. There are other things that you may think of later.

Look out the windows and see if there are any things you can see in the air. There may be clouds—or a bird may swoop past—you may be lucky enough to see an airplane in the distance—a fly or some other insect may be fluttering against the window. There will probably be something flying or floating through the air. The sky isn't just an empty space, is it?

Explain to the children that they are going to make collages of things they might expect to see in the sky. You won't want to include everything that might be in the air; that would be much too crowded. You will make a picture that says these are things that are sky things—they belong there.

Show the class the materials that are available. Include as much variety of paper as possible: construction paper, corrugated paper, sandpaper, and printed paper (gift wrapping and wallpaper). If you have thin sheets of cork, include them. You won't want to include all of them, but choose several kinds so your picture will have a variety of textures. Show them that there are two shapes of background paper—the regular 12″ × 18″ size and a long, narrow kind 9″ × 24″—so you will also have to decide what shape your collage will be. Either size paper can be used either vertically or horizontally.

There will be other questions, too, so take time to answer them. You won't want them to be all birds or all airplanes—or any one thing. Your collage is to say that there are many things in the air, so you will want to show many of them. You will make each thing that is going to be a part of your collage, and then you will arrange them so they look natural. They can be any color you like, but you will have to choose both color and texture to make a pleasing collage. Remember, too, that when you arrange things in a collage it is a good idea to do some overlapping. Everything doesn't have to overlap, but group things so they look as though they belong together.

Let each child choose the size and color background paper he would like and a few pieces of small paper. While this is being done, see that each child has a pair of scissors. Those will be the only materials he will need for a few minutes. While you go about the room to assist each child, see that he gets pasting supplies—a bit of paste on a scrap paper, a paste brush, and a newspaper on which to do the pasting.

Good! I'm glad you're making something large first. Oh, but that's too tiny for a collage mosquito. It would never be seen. Sure you can use it—just make it bigger. You won't be able to use blue paper on the same kind of blue background, will you? It wouldn't show—and that part of your picture would be lost. Oh, yes—that will be much better. You have a wonderful idea that no one else thought of! Of course you would see stars in the sky—and a sun or moon, too! That was good thinking. Why don't you paste the wings and head and spots on your beetle. Then you could move him more easily until you find just the right place for him on the collage. That's an excellent repeating of color. Do you have one area that looks empty? Well, it's easy enough to make something that will fit into it. You are ready to paste everything to the background, aren't you?

As each child finishes his picture, have him clear away all extra materials. Then encourage him to think of a suitable title and put it on the back of his collage, so you can use it in the display.

Before that time, of course, you will want each child to show his work to the rest of the class. Have him give the title and make any explanations he wants. Encourage the other children to notice unusual ideas, arrangement of the parts that create effective eye movement through the picture, interesting choice of colors and textures, titles that improve the pictures—anything that shows good thinking and good workmanship. The whole collage doesn't have to be successful for some part of it to be worth recognition—and to make the child feel that his efforts were worthwhile.

You'll never again think of the sky as empty—there's always something up in the air. The display in your room will prove it.

Make It Easy—For Yourself!

1. No pencils! Children will do better if they are taught not to do any preliminary drawing. Think and then cut.
2. If each child takes only a couple of the materials at first, it will

permit all the class to begin work quickly. Encourage them to share leftover supplies with other children. Of course, let them return to the supply area when different or larger materials than they can get from other children are needed.

3. If "collage" is a new word, explain that it is a French word that means pasting. A collage frequently has a variety of textures or materials in it.

4. Have a variety of 9" × 12" colored construction paper. The other textured materials can be cut to a somewhat smaller size to stretch the supply.

5. Encourage the children to repeat the same colors and textures in different parts of the collages. This will help add rhythm and balance to the completed pictures.

6. Encourage children to move and rearrange parts of their collages until just the right effect is created before they paste anything to the background.

7. Teach the children to do their pasting on newspaper in order to keep their pictures and work areas clean. Paste around all the edges of an object—not in the center, or with dots of paste.

8. If paste brushes are not available, paste applicators can be made by folding scraps of paper until they are narrow strips, then bending them in the middle for added strength.

9. A creative title can make an excellent picture out of an otherwise ordinary one. Put the title on the back where it is available for display purposes but won't interfere with the appearance of the picture.

I am roller-skating with my friend.
There are big clouds, but the sun is out.

lesson 2
Cloudy Weather

OBJECTIVES

1. To be more aware of the kinds of cloud formations.
2. To experiment with an unusual combination of materials.
3. To express an idea realistically.

The weather report is for cloudy weather. But that can mean almost anything!

Choose a day when there are clouds in the sky—big, puffy, white ones that float lazily along; such thick ones that they cover the whole sky with a gray shroud; mountainous thunderheads that boil ominously—any kind of clouds.

Have you ever noticed how often there are clouds in the sky? Just look at those clouds up there today!

Talk about them with your class. Comment about the clear blue sky around them if it is a sunny day—or the grayness of the whole sky—or whatever is appropriate. Notice the shape of the clouds, the color of them. Do they seem to be moving fast or do they appear to stay still?

Do clouds always look like these? Do you see clouds on a sunny day as well as when there is going to be a storm? How do they look different? If you were out riding your bicycle and you saw some clouds in the sky, could you tell if you needed to hurry home or if it would be all right to continue riding? How could you tell?

Encourage all the children to add something to the conversation. Talk about color, shape, and movement of the clouds. Some children will have had experiences they want to tell you about. It was a good thing you had an umbrella with you, wasn't it? Ask a question that will make the experience more visual.

Hold up a piece of cotton batting. Could you make some clouds out of this? It does look something like a cloud already, doesn't it? But if you were going to make a cloud picture with it, you would have to pull it apart a little so it wouldn't be so thick; then it would stick to your paper better. The first thing you will have to decide, though, is what kind of clouds are going to be in your picture. If it is going to be raining, will your clouds look like clouds that you would see on a sunny day? Not at all! Would thunderstorm clouds look like the kind you see when it rains all day?

You decide what you are going to put in your picture so that later you can draw all those things with crayon. It can be a summer picture, or it can be a winter picture, if you like. Of course it doesn't have to be raining or snowing; there are clouds even when the sun is shining.

Explain that they will make the clouds first. Paint the shape of the clouds with paste and then lay the cotton batting on top of the paste. You may then use your crayons to draw the rest of the picture—the part that will tell us what is happening on that particular cloudy day.

Give each child a piece of 12″ × 18″ white drawing paper, a bit of paste on a scrap paper, a paste brush, and a small handful of cotton batting. Walk about the room, commenting about the different kinds of clouds you see being made.

Um-m, the sun must be out with those tiny, puffy clouds. You will have lots of blue sky showing, won't you? I wonder if they are going to be thunder clouds. No, don't tell me! I'll come back later when you have the rest of the picture drawn with crayon. Use your crayons to finish the picture as soon as you have made the clouds. I thought those were rain clouds, and now I am sure of it!

When you see that children have a good beginning with the crayon part of their pictures, cover a table or group of desks (or even a space on the floor) with newspaper. Put some blue paint, some black paint, and some brown paint in the center of it, and have two or three brushes for each color of paint. Then ask the class to stop work for a few moments.

Tell them the three colors of paint that are available. Which one would help your clouds look like sunny day clouds? The blue would, of course. Which one would help make the stormiest clouds? The black would do that—and the brown might make just light rain or snow clouds. Which kind of clouds do you have on your picture?

Have your class gather around you at the painting area. Have a piece of paper on which you have previously pasted some cotton batting clouds. There isn't any crayon picture on here, so we will just have to look at the shape of the clouds to tell what kind of clouds they are—what kind of day it is. What color paint do you think we should use on them? Blue? Yes, they could be big, fluffy, sunny weather clouds.

Dip an easel brush into the blue paint, and wipe off the excess paint so it doesn't drip. Then lay the side of the brush flat on the cotton batting cloud. Pat it up and down where you want the color. The whole cloud won't be blue, will it? It is a white cloud and the paint is just to make some shadow on it—and to help say that it is a pleasant weather cloud; the paint just lays on top. Call the children's attention to the fact that you are not brushing the paint on as they usually do when they paint. They will just pat the paint on with the side of the brush. No, you wouldn't use blue if you have a rainy day picture.

Let groups of children take turns at the painting area. If the crayon part of the picture isn't finished, they can complete it either before or after adding a bit of paint to the clouds. You won't put any paint on the paper—just crayon and cotton batting—and a little paint on the cotton batting clouds.

Supervise the painting area until each child has had an opportunity to add paint to his clouds. Answer any questions or help a child who may have a problem with the technique. You would only want to use one color on your clouds—the one which tells us best what kind they are. Yes, layers of clouds like those are darker underneath, aren't they! That's a fine way to do it.

Give each child a chance to show his picture to the rest of the class and to make any comments about it. Later display all the work—perhaps grouping them by the kind of day that is shown.

Cloudy weather isn't always bad weather, is it? And even bad weather can be good when all you have to do is look at it!

Make It Easy—For Yourself!

1. Older children may know the names of various kinds of clouds. If they do, let them use their correct names—cirrus, nimbus, stratus, cumulus. Some children may want to use encyclopedias or science books to find information about them; the librarian will be glad to help.

2. If paste brushes are not available, have the children make paste

applicators. Fold a scrap of paper until it is a narrow strip, then bend it in the middle to give it added strength. Keep fingers out of paste—paste and cotton batting make furry fingers that are not conducive to good art work!

3. Egg cartons make fine palettes. An egg carton broken in half will provide ample space for all three colors; very little paint is needed.
4. As children finish with the cotton batting, have them put any extra cotton back with the supplies. They should also put back any extra paste and throw away the scrap paper. Paste brushes should be left, to be washed later. If you use nylon paste brushes, they can be left to dry with the paste on them. When they are thoroughly dry the dry paste will come off easily by running a finger across the tops of them. (Chunks of paste should always be wiped off, of course.)
5. Stack the egg carton palettes, wrap them in the newspaper that covered the painting area, and discard the bundle.

9 Weather

Continued Rain

lesson 1
Forecast . . . . . .

OBJECTIVES

1. To relate the effect of weather to activities.
2. To associate color with mood.
3. To provide opportunity for a cut paper activity.
4. To experiment with fitting a picture to an unusual shape.

One person says it's going to rain; another says no, it's going to be sunny. Neither one is right—it's going to snow. You know because you can forecast the weather.

How would you like to make just the kind of weather you like? It would be great, wouldn't it? What kind of weather would you make?

Someone will almost certainly want it to be sunny. It's wonderful to have the sun shine, and if you could have it sunny any time you liked, you'd never have any picnics spoiled, would you? What else do you like to do when the sun is shining? When it's sunny in the summer you like to go swimming. If it's at the beach, it's nice to lie on the warm sand—or maybe to dig into the wet sand. You like to play baseball with your friends. What are some other things you like to do outdoors on a warm, sunny day?

Many things will be suggested—roller skating, riding a bicycle, flying kites, hiking, climbing trees, jumping rope, playing hopscotch. Each time, ask a question or add a comment of your own that will add to the visual effect and increase the children's thinking. Apple trees are better to climb than oak trees, aren't they? The low, twisting branches give you an easy start. Do you jump rope by yourself, or do you have two friends to turn the rope for you? Do things look dull or bright on a sunny day? The sun makes everything brighter. Sometimes it makes things so bright you have to wear sunglasses.

Do you think you would ever like to have it rain? It wouldn't be good to have it rain while you were at the beach, but it is important to have it rain sometimes, isn't it? Flowers and trees and all kinds of living things need it to help them grow. What do you like to do on rainy days? Do you ever go out in the rain? What do you look like in the rain? Are the colors as bright as on a sunny day?

Again take time to talk, to question, to comment—to get a variety of ideas and to make things visual. Then switch to snow; encourage each child to make some contribution to the discussion.

Let's make a picture about the weather—any kind of weather you would like to have. If it's rain you want, you might be splashing in that puddle you told us about. If you would like to have a snowstorm so you can help your daddy shovel out the driveway, you may make a picture of that. You might be roller-skating on the sidewalk in front of your house if you have the sun shining, or you can make any one of the other good ideas you had. Just choose which one you would like to make into a picture.

Show the class the variety of construction paper. All the parts of your pictures will be made out of white or colored paper. You will have to think of what the weather is like and choose colors that will tell us that it is a dark, rainy or snowy day—or a bright, sunny day.

There's just one other thing you'll have to do first—you'll have to make the rain or snow or sun. There won't have to be a sun *in* the picture —the picture will be *in the sun*, or it might be in the rain or in the snow. How can that be? Easy. Take a dark or dull color paper and begin to cut it as you talk. Let's see—if I started at the middle, at the top of the paper, and cut in a slightly curved line out to the edge of the paper—and then made a circular cut like this around the bottom of the paper—and then the same way on the other side—and back up in a point to where I started from. See—a big raindrop, big enough to make a rainy day picture right on top of it. You could put some rain in the picture, too, if you wanted to, but you wouldn't have to. Just the shape of the paper—a big drop of rain—would tell us what the weather was like. The picture you pasted on top of it would tell us, too.

What would you do if you were going to make a sunny day picture? You would make a big sun first—as big as you could make it out of one of these pieces of 12″ × 18″ paper. It would be easy to make a circle, wouldn't it? Could you do anything else so people would be sure it was a sun? Yes, you could use some of the leftover paper to make pointed rays of sun; then it would be really shining brightly.

How would you make a shape for a snow picture? Someone will certainly suggest a snowflake. If you have younger children, let them make a very simple four-pointed star. If you have older children, you may want to show them how to get six points to their snowflakes. A snowflake is lacy, though, with open designs in it. How could you make it look like a snowflake instead of just a star? You could cut shapes into it, but you would have to be careful not to cut too many, because you will need plenty of space for your picture. You could cut the designs in the points only and leave the rest of the snowflake for the picture.

Well, you decide what kind of weather you would like—and then you make that weather—a big sun, a lacy snowflake, or a giant raindrop. Then you plan your picture for that kind of weather, remembering to choose colors that belong to the weather you have.

Let groups of children take turns selecting the colors for their backgrounds. See that each child has a pair of scissors so that he can begin immediately.

Make it as big as you can. Your sun can be as wide as your paper is, can't it? Can you round it a little more than that? Go all the way to the side—and all the way to the bottom for the raindrop. It won't do any harm if you didn't start quite in the center. Oh, I think you can cut more smoothly than that! See—it is better already. Push the blade of the scissors through the points of your snowflake to cut out designs. If you folded your paper to cut the star, you can leave it folded to cut out the shapes, but be sure to cut on the folded edge. Orange points on your yellow sun? Of course, that might make it an even brighter sun.

The children making a sunny day picture will need paste almost immediately so they can paste the sun rays to the circle. You can give them a paste brush and a bit of paste on a scrap paper as you see who needs it.

Begin the picture part of your weather as soon as you are ready for it. Choose only a few colors to begin with, and when you need only a small amount of other colors perhaps you can find someone who has some little pieces he will let you use. Be sure what you make will fit the sun you have. You will paste it to the sun as soon as you are sure it is arranged just the way you want it. Could you tip the umbrella a bit so all of it will fit on the raindrop? It makes it look as though he is holding it against the wind, doesn't it?

Continue to help each child in any way he needs until everyone has achieved his best work. A word of approval, a question, a compliment, or a comment will keep children thinking and interested until their pictures are finished; then they will want to show them to the rest of the class—and to tell something about them. Encourage the rest of the class to comment about the unusual ideas, the clever way in which parts have been arranged to fit the background shape, the appropriateness of colors that were used, good workmanship, or some unique touch. Don't expect any child to have done each thing perfectly, but do acknowledge something that is good about each child's picture.

Look! It's raining, it's snowing, and the sun is out, all at the same time! How would you ever forecast that kind of weather?

Make It Easy—For Yourself!

1. Have 12″ × 18″ construction paper for the raindrops, snowflakes, and suns. Cut 9″ × 12″ white and colored construction paper in half for the picture part. Have enough choice of colors so that children will be able to choose colors that are appropriate to the weather.
2. No pencils! Children will do better if they think and cut without any preliminary drawing. Don't let them add any pencil or crayon details to their pictures.
3. Don't be in a hurry to give out paste (except to those children who are making a sun). Encourage them to make parts and re-arrange them before they paste them to the background. This gives you an opportunity to help them, as well as to teach them to plan their pictures.
4. Give each child a small piece of newspaper—half a page is large enough—to do his pasting on. It may be folded in half to take less space on the desk, and it can be folded in reverse when more clean area is needed.
5. If paste brushes are not available, have each child make his own paste applicator by folding a scrap of paper until it is a narrow strip, bending it in the middle to give it added strength. Teach children to keep their fingers out of the paste!
6. As you walk about the room, carry a 12″ × 18″ paper as a tray on which children may pile flat scraps of paper. This will give them additional work space on their desks and will make the final cleanup easier and faster.

On the Loose

lesson 2
Blowing Up a Storm!

OBJECTIVES

1. To make a visual interpretation of a storm.
2. To experiment with color and line to interpret sound and motion.
3. To have experience with a material which makes a rapid expression possible.

There's an ominous feeling in the air. The sky is filled with black and angry clouds. It's blowing up a storm!

Have you ever been away and had to hurry home because there was a storm coming?

Encourage children to tell briefly of some experience. Yes, a storm could ruin a picnic, couldn't it? Did you have time to cook those hotdogs, or did you have to take them into the house to cook them? It must have been frightening to have the lightning strike the tree! Did it do any damage? Have you ever been on the water during a storm? I'm sure a storm would seem even worse on water than on land. What were the waves like? Like mountains of water! Could you feel the spray as the wind blew the water? Have you ever been in a hurricane? Or a tornado? Or a blizzard?

Talk about the characteristics of each kind of storm. What did the sky look like? What did it feel like? Wind is an important part of most

storms. How could you show wind in a picture? If the trees were bending, it would show that the wind was blowing them; lines on a picture
could show the direction of wind. How else could you show that there
was a strong wind?

Then bring the conversation around to color. What is a good storm
color? The sky might become almost black with heavy clouds. Black is a
strong, heavy color, so it would make a storm look heavy, too. Would you
use light colors for a storm? No, a gentle rain might be done with light
colors, but not a severe storm. Bright, sharp colors would be fine for
wind or lightning. They would make the wind feel sharp—and the lightning look sharp. Heavy or sharp colors are loud so they would be good for
thunder, too. When you make a picture of a storm, you will need to think
of colors which help you feel and see the storm. What kinds of lines are
dangerous lines for a dangerous storm? Sharp lines are severe lines—
they're frightening lines; they would help create the mood of the storm.

Explain to the children that they will use colored chalk to make a
picture of some kind of storm. It may be a hurricane, blizzard, tornado,
thunderstorm, or just a severe rain and wind storm. It will be realistic—
or at least some part of it will be realistic. Your picture will make us see
the storm, but it will also make us feel it—and hear it. It will frighten us
by putting us in the midst of the storm and sensing the danger of it. The
colors you use will help you do that; so will the lines. Are you ready?
Have you thought of a good idea?

Give each child a piece of 12″ × 18″ white drawing paper and a box
of colored chalk. Remind them to plan their pictures before they begin.

That has to be a tornado with that kind of funnel-shaped cloud,
doesn't it? There will be damage where it touches the ground; I'll have
to see later what happens. No, you don't have to be out in the storm. You
would be foolish to stay in it if you could get out of it; just show us the
storm, and make us hear and feel its fury. Does this part of your picture
look too quiet—perhaps a little empty? Can you think of something to
put there? There can't be any sun in any of these pictures, can there?
It wouldn't be much of a storm, would it! I'm glad you made the boat
small; it makes it seem even more frightening that way. Can you think
of any way of making sharp lines on the water? That's a good idea! The
wind would do that to them.

As the children begin to finish their pictures, urge them to accent
some parts of them. Just press harder to make a sharper, more important
line. See how it makes that part of the picture more important! Try doing
that on some other parts of your picture. Do you have a good title for your
picture?

When all the children have finished, have a sharing time when each
child can show his picture and tell the title or anything important about

it, or let other children in the class describe the storm before the person who made the picture tells about it. Later display all of the pictures.

Aren't you glad you're indoors where it's dry and secure? With all those storms blowing up, it wouldn't be safe to venture out into them.

Make It Easy—For Yourself!

1. Cover all work areas with newspaper, since this will cut down on the cleanup problem.
2. If you are using new boxes of chalk for the first time, have each child break each piece of chalk in the box in half before he uses it. This does not spoil the chalk—it protects it as well as makes it easier to use. Half sticks of chalk are less likely than whole sticks to break into tiny pieces that have to be discarded. In addition, it doubles your supply.
3. No pencils! No preliminary drawing. Think and then draw with chalk.
4. Encourage children to draw masses of color by using the side of the chalk. Sharp lines and accents should be drawn with the point.
5. Hold the chalk close to the paper when drawing with the end of it. Tiny pieces will still break off, but larger pieces will not.
6. The paper will absorb only a limited amount of chalk; the rest will remain loose as colored dust. Teach the children to occasionally drop it off onto the newspaper or floor—or to gently blow it off, taking care not to bother anyone else.
7. Give each child a piece of facial tissue so he can soften some chalk areas in order to create a stronger accent with a neighboring area. Wrap the tissue around the index finger and rub softly.
8. If the pictures are to be displayed where fingers or clothing might rub against them, spray them with a light coat of fixative. Ordinary hair spray is a fine substitute for a fixative.
9. Urge each child to think of a creative title which will add to the mood of his picture. A good title can improve a picture. Titles should be put on the back of the papers where they won't detract from the pictures, but where they are available for use when the pictures are displayed.
10. Older children may want to use the library to find additional information about particular kinds of storms. Display their reports with the chalk pictures.

10
Air and Wind

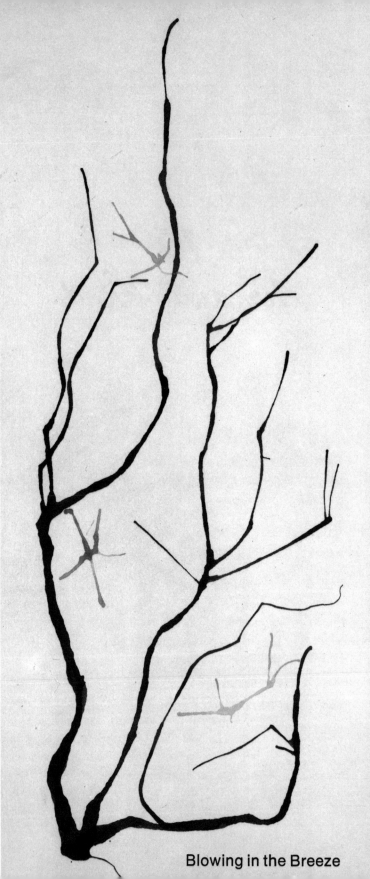

Blowing in the Breeze

lesson 1
Blow, Blow, Blow

OBJECTIVES

1. To demonstrate the moving force of air.
2. To experiment with an unusual painting technique.
3. To appreciate the delicacy of line painting.
4. To be aware of the possibilities of a partly accidental, partly controlled painting.

You hear it; you feel it; you see it. The wind never stops—there's a constant blow, blow, blow.

Were you ever out in such a strong wind that you couldn't stand still—it pushed you along? Take only a minute or two for children to comment about it.

You're not going to be blown by the wind today—you are going to be the wind, and you are going to blow something else. No—you're not going to blow people—you're going to blow paint!

Have your class gather around to watch while you demonstrate. You will need a piece of 12″ × 18″ white drawing paper, some thin black paint, a tongue depressor, and a drinking straw. Dip the tongue depressor into the paint and let it drip onto the white paper. Place the spot of paint near the bottom of the paper and slightly to one side of center. Add two to four more drops of paint on the same spot, but don't let the tongue depressor touch the paper.

As you do this, talk to the class. I said we are going to blow paint, but if it dries first, not even the strongest wind could make it move, could it? The paper would blow away, but the paint wouldn't move on it. I don't want to blow the paper—just the paint, so I will have to work fast.

Place a drinking straw in your mouth so you can blow through it. Kneel or bend down low so the straw will be almost parallel with the paper. Place the end of the straw just below the puddle of paint, but don't let the straw touch the paint. Take a deep breath and blow through the straw slowly and continuously. A line of paint will begin to travel up the paper; follow the straw close behind it to keep the paint moving. Branches of paint will form; follow a big one as far as it will go. Take another deep breath and push along another branch of paint. Start with the same puddle again and move any remaining paint in another line. As long as there is any remaining wet paint, move it.

My goodness! I guess that looks easy, but it really takes a lot of breath! It's hard work making that much wind.

Look what has happened to that puddle of paint that was near the bottom of the paper. It isn't even a puddle anymore; it has been blown into thin, graceful, tree-like lines that have moved up and out, and have gotten thinner and thinner until nothing is left. Did you notice that I aimed the straw in different directions to make the paint move in the direction I wanted it to go? When that line got close to the edge of the paper, I changed the direction of the wind. Well, regular wind changes directions sometimes, too. Have you ever seen a paper or a leaf blowing— and suddenly it changed direction and blew a different way?

Did you notice that when all the paint on one branch was gone, I moved quickly to another branch where there was some wet paint? Even a tiny branch may have a speck of wet paint at the end of it. As long as there is any left, move it just as far as it will go.

Why do you suppose I had the straw almost flat? If the straw was straight up and down the wind would blow the paint flat against the paper—all it could do would be spread out in all directions. The strongest wind blows things across. Will you be able to sit while you blow the paint? No, that would make the wind blow down. You will have to get down on the floor so you can make the wind blow straight across the paint.

Everyone will be eager to start—and sure that it is easy.

Lay your paper the tall way, so you will have plenty of space for the paint to blow. Remember to leave an inch or two of paper below the puddle of paint. Yes, that's plenty of paint. Oh, get down lower than that— so the straw will be almost flat. Take a deep breath—and blow gradually and continuously. Don't stop! Keep blowing—keep that paint moving. Don't let the straw touch the paint. You have more paint there; blow it in a longer line. Can you make it move in a different direction? There is more paint in the puddle; see if you can start another branch. See—there's one started. Of course it is harder to do than it looked! It takes a lot of breath to make a big wind! It won't do any good to blow after all the paint has been used.

While children recover from their blowing, move about the room and hold up a picture here and there for everyone to see. Isn't this a tall picture! Did you get your paint to move up as high on your paper as he did on this one? Look at the double line on this painting. Did anyone else have that happen? Yours didn't move very far, but it made lots of delicate branches, didn't it? We can trim it with the paper cutter so it will fit the paper; yours will look nicer if we do that, too.

Can you blow just a little more? This will be just a tiny gust of wind —not like the gale you blew before.

Have the class again gather around you to watch. This time I'm going to add just a speck of one other color to this picture. I'll drop just the tiniest bit of color down here near the bottom of my picture. See, there's an empty space where no black lines blew out from the puddle.

As you talk, touch a tongue depressor to the paint and drop the smallest amount of paint possible; it isn't much, but it is plenty. This time I'm going to hold the straw right above that spot of paint, and I'm going to blow straight down on it with one quick blow. What do you think will happen? That's just what should happen! Let's see if it does. Give one quick puff through the straw—close to the paint and exactly above it. See! You were right! It did splatter out in all directions. It was like a little explosion!

Where would be a good place to add another little explosion of this same color? Well—do you think that would be too far away from the black lines? Would it be better if I put it over here, so it looks as though it belongs to that long line? My straw was tipped slightly so more of the explosion is on one side of the paint. Yes, they do look like flowers; let's make one more flower explode into a blossom. Let a child choose a place for it, then you drop a speck of paint, and blow it from above. Yes, it is nice—and yours will be, too. Ready?

Give each child a tiny bit of whatever color he chooses. This part of the painting will be done so quickly that some children will have finished before you have given paint to other children, so while you dispense the paint, constantly remind the children to use the tiniest bit of paint possible; too much paint will spoil the effectiveness of it. Hold that straw right above the paint; one quick puff is enough.

Collect and dispose of the straws and extra paint, and then let each child show his work to the rest of the class. They will be proud of them—and you will be, too.

Everyone who sees them will want to know how you could get such fine lines. Just pretend you are the wind—and blow and blow and blow. You blow almost as hard as the wind.

Make It Easy—For Yourself!

1. Thin the tempera paint until it is the consistency of ink—or use ink in place of paint. It must be thin to move in long, graceful lines. Try using it yourself before the class uses it.
2. Egg cartons make fine paint palettes. Each carton can be broken into six parts, so each child will have two sections to his palette. Put a small amount of black into one section; the other section can be used later for the second color. Don't give out the second colors until you have demonstrated with one of them.
3. Straws must be held almost flat to make the air move in long lines. Do exactly the opposite for the spots of color—hold the straws vertically to make the color explode in all directions.
4. Urge the children not to blow too long. Have them stop before they become too tired or out of breath.
5. Very young children may not be able to blow the long lines. Let them use several spots of two or three colors and blow down on them; a couple of drops should be enough for each spot.

Look Up!

lesson 2
Windy Weather

OBJECTIVES

1. To be more aware of the effect of wind and air.
2. To provide opportunity for participating in a group activity.
3. To experience working with a fluid media that makes a rapid expression possible.

Windy weather is wonderful weather!

Have you ever been out on such a windy day that you felt like the wind was going to pick you up and blow you away? Well, it might feel that way, but you knew it couldn't happen, didn't you? The wind does pick up other things, though; the wind would blow scraps of paper if someone was careless and left them around, but they shouldn't be there, should they? Let's think of things you expect to see.

Kites, yes! That's a good beginning. Someone else will probably suggest balloons; perhaps clouds. Can you think of some things that can fly because the air holds them up? Birds of all kinds are helped to stay up in the air: robins, sparrows, mockingbirds, blue jays, seagulls, eagles, hawks, owls—all kinds of birds. What are some other living things that can stay up in the air? Suggestions will include butterflies, mosquitoes, flies, bees, moths, dragonflies, beetles, wasps, and bugs of all kinds.

That's a long list of things already, but there is another type of thing that the air helps hold up and move along. Hornets can stay up in the air, too, but they are living things like the wasps and bugs we talked about. I am thinking of some things that aren't alive, but they have engines that help them move along as the air helps hold them up. Certainly—airplanes; all kinds of airplanes, from tiny ones to jumbo jets. There are rockets and blimps—and gliders, too. Oh, yes—helicopters are a good idea.

That's such a long list of things the air holds up that it will take a giant picture for all of them. We're going to put all of them on one picture —but just one picture for everybody. That's right! It will be a mural.

Unroll a long piece of mural paper on the floor. Have the children place their chairs in a row on each side of the paper, but keep them far enough away from it so there will be enough space between them and the paper for the children to paint.

What would be a good thing to make first—something that you would like to make big? A jet is a fine idea because that really is big. It would be perfectly all right to make a bumblebee first and make him a giant one; things don't have to be their right size on this mural. Of course they don't have to be! A jet its right size would be many, many times too big to even fit on the whole paper—and a mosquito its right size would be so small that nobody would see him at all. How about beginning the mural with both of those things—and making both of them big?

Let the two children who suggested them be the ones to begin the mural. Have each of them get an easel brush and half an egg carton for a palette. Give each child the first color paint he needs, and let him go to work. Remember to make them big because that is the reason they are being painted first. It is all right for you both to work on opposite sides of the mural if you want to. An artist might make some things upside down in a mural; do it whichever way you like.

Have all the other children watch for a minute or two until you are sure the two beginning children get a good start. Then choose two or three more children who want a large space for what they would like to make. Let them get brushes and egg cartons, and give them each the color paint they need.

Again let the other children watch while you have an opportunity to see what is going on.

You are certainly making a big jet. Would you like to let someone else help finish it now that you have made such a good beginning? Fine—choose someone to take your place. Get some more paint if you need it. I should think you would want another color to add to your bumblebee.

Leave the brush that has the paint on it on the sharing desk; maybe someone else will use it later for the same color paint. You get a clean brush for the new color and I'll add some of the second color to your palette.

Continue to let other children add to the mural. Be sure something different is added each time and make sure each child finds a suitable work space—one where there is room for him to work and where there is a good place for the size and color picture he is going to paint.

When a child needs another color paint, have him bring his palette to the supply area so you can give him some of the new color. Have him leave the brush there so someone else may use it in the same color. If there is another brush which has been in the new color he is going to use, he can take that; if none is available let him take a clean brush. Egg carton palettes may be left at the sharing area and reused by another child if they contain a color he needs. In this way fewer supplies will be used, so less cleanup will be necessary. It will also mean that cans of water for washing brushes will not be needed—so the possibility of water being spilled on the mural will be eliminated.

Be sure each child paints one thing on the mural; then let children paint a second time. As the spaces become smaller, have the children think of less important things that can be put into them.

No, it doesn't have to be windy weather for all those things to stay in the air—but it will help some of them, won't it?

Make It Easy—For Yourself!

1. Cut the mural paper long enough so each child will be able to paint on it at least once, or cut it to fit a specific space—as long as all the children will be able to work on it. Push as many desks as necessary out of the way to provide enough space on the floor for the mural paper and for the class to sit around it.
2. Egg cartons make satisfactory palettes. Break them in half to hold up to six colors.
3. Children should hold their palettes while they are working. This allows them to hold the palette close to the area they are painting and so eliminate drips across the mural; it also prevents the palettes from being tipped over by other children knocking against them.
4. Remind children to wipe their brushes once inside the palette to prevent paint from dripping.

5. Encourage children to hold the brushes at the top of the handles to provide more freedom of motion. When children hold brushes as they do pencils or crayons, their control is limited and the results tend to be tiny and insignificant.

6. Re-use the brushes and palettes. As each child finishes painting, have him leave his brush and palette on a desk or table that has been covered with newspaper. When another child is ready to paint, let him choose a brush and palette that has been used for the same color he wants to use. Add any colors that are needed, and get a clean brush when a new color is to be used.

7. There will be almost no cleanup. Leave the brushes on a newspaper at the sink so they can be washed later. Lightly stack the cartons, wrap them in the newspaper that covered the desk, and discard them.

8. If it is possible, leave the mural on the floor to dry so the paint won't run or be smudged when it is tacked to the bulletin board.

part three
Water

11 Animals of the Sea

lesson 1

Down to the Sea

OBJECTIVES

1. To concentrate attention on the variety of sea life.
2. To experiment with two similar materials used in different ways.
3. To introduce sketching with charcoal.
4. To encourage children to work loosely and freely.

No, you don't have to have a ship to go down to the sea—but it would be nice if you had a glass-bottomed boat. Think of all the things you'd see!

Have you ever stood on the beach and watched the water spread up on the sand? If it was a calm day, the water rolled in gently and then moved back again. Did the waves move in straight lines? They made slowly rolling lines—not straight lines and not wiggly lines. Move your hand in slow, wave-like lines. You can almost feel the motion of the water, can't you?

What color is the sea? Just plain blue? There are shadows of other colors; you would see some green in the water. Blue and green are cool colors—even purple. The water is cool, too. You wouldn't find any red or orange in the water, for they are hot colors—sun colors.

In the Deep

Suppose you could go deep down into the water. What kinds of things do you think you might find swimming there? There will be fish— big fish, little fish, all kinds of fish. There will be regular-shaped fish like trout and salmon and bluefish, but there will be fish that are shaped differently, too. Yes, a whale is not only big, but there is something about the shape of him that makes him look different from other fish. His tail is small compared to his body, and his tail is flat horizontally. What are some other fish that are a little different? How is a dolphin different from other fish? Well, he has a long nose and a large but thin body. Did you know that some dolphin are called porpoises? A swordfish has something about him that makes him different from fish like the codfish. The top part of his mouth is a long, swordlike weapon. The sailfish, also, has a swordlike upper part to his mouth, but he has a large upper fin that looks like a sail. Sometimes they can glide through the air for a while after leaping from the water. We'd call them flying fish, wouldn't we?

Can you think of other things beside fish that move through the water? There are lobsters—and other things which we call shellfish. Can you think of some of them? Crabs, oysters, clams, mussels, shrimp. Take time to describe some particular characteristic of each one—the long tail and long, curved antenna of the shrimp; the short, broad body of the crab; the irregularly-shaped shell of the oyster; the variety of sizes and shapes of clam shells; the long, narrow oval of the mussel.

All those things had shells. There are some kinds of sea life that have soft bodies—sometimes they even change shape in the winter. You're right —jellyfish! There are all kinds of jellyfish; some have bodies like umbrellas —with long, trailing tentacles. Yes, some of them do sting if they touch you. No, they aren't stingrays; a stingray stings you, too, but he looks different. He has an almost diamond-shaped body with a long, flexible tail.

What are some other soft fish? An octopus has a soft, oval body with long arms that reach out to suck in food. Do you know how many arms an octopus has? There are eight arms to an octopus, who lives mostly on the bottom of the sea. A squid has arms that reach out—ten of them—but it has a long shell. Did you know that a squid can move backwards faster than it can go forward? It also ejects an inky fluid when it is in danger.

There are other kinds of sea life and perhaps you will be able to think of some, but let's see what we are going to do with these things.

Have your class gather around you. Lay a piece of 12″ × 18″ white drawing paper on a piece of newspaper, and ask two children to choose a good color for water from the chalk you have. Fine! Now let's see what we can do with them.

Lay one piece of chalk flat on the paper. Water moves gently and easily—press on the chalk and move it in long, rolling lines. Every wave doesn't come to the shore in exactly the same shape or size, so I'll move the chalk to another area of the paper and make some more motions with it—a long one like that, and a shorter, flatter one like that.

When more than half the paper is covered with the first color, do the same with the second color. Fill all the paper with color and let the two of them overlap in some places. That only takes a minute. See all the motion there is to the color—to the water. Oh, no, it isn't rough enough to make you seasick! Just a nice, gentle, rolling motion. I could lighten some parts of it by rubbing it softly with this facial tissue wrapped around my finger—or I could make some areas stronger by going over it with the chalk again and pressing a little harder on it. But the sea isn't an empty place like this, so let's do something about that.

I'll use charcoal to sketch an ordinary fish, but I'll make it large since it is the first thing. Would it be a good idea to put it right in the middle of the picture? No, that would be uninteresting because the rest of the space would be the same size. It will be much better to put it off-center—like that. Sketch a large fish as you talk. Yes, I can give it an eye and a fin or two.

Where would be a good place to put something else? There is plenty of space for something big there. What would you like to have there? A squid? There's that long, almost straight body—and head—and tail. Now it needs ten tentacles reaching out—and two long, curved antennae.

The paper isn't nearly filled, but let's pretend there is an oyster over there, and a sailfish there—and there would still be space for several more things. Did you notice that when I sketched I moved the charcoal in quick, loose motions? They were almost like the motion of the water; each time the direction of the motion changed, I lifted the charcoal off the paper. Make another smaller fish as you talk—one motion for the top of the fish, another for the bottom. If its mouth is open, add two more motion lines. Do the same thing for the tail—and a fin. The charcoal just moves in the motion of the shape. The lines are light, though, because I sketched lightly and freely. Now I'll go over the lines again to make them darker—some bits of charcoal will break off, but they will be little pieces because I'm holding the charcoal so close to the end that I'm using.

Let's not finish this picture; instead, you will make yours. Give each child the materials he will need for the background color.

Two colors are enough—you just want the color to make a cool, motion background. That's fine! It really looks like it is moving. Don't you need more of the first color? Save some areas of the paper for the second color. No, don't draw lines for the background. Lay the chalk flat on the paper and rub it in the motion of water. See—isn't that better? Use your second color whenever you are ready for it.

As you comment to the children, give each one a piece of charcoal so he will have it any time he is ready to use it.

That is a wonderful giant lobster. Do you like to eat lobster? That must be why you gave him such big claws! Can you think of something different to add to your picture that no one else has thought of? Good! You are the only one who thought of an eel. He really is moving along easily! Does anyone have anything else different in your picture? Of course! We forgot all about starfish! I'm glad you remembered it. Make each part of it a separate motion. That's a beautiful sailfish! Did you ever see a real one? You must have seen lots of pictures of them, then. You are ready to start making yours darker, aren't you? Oh, press harder than that! Don't worry if tiny pieces of charcoal break off. That's the way to do it.

The children will be pleased with their finished pictures—and so will you. Give them time to show them to the rest of the class. Let them—and the other children, too—talk about the variety of things they have sketched. Comment about the feeling of motion in the background, the good sketching technique. Arrange for all of them to be displayed.

All that water makes you feel as though you had gone down to the sea—in a glass-bottomed boat! Well, you wouldn't see any greater variety of creatures if you had!

Make It Easy—For Yourself!

1. Cover all work areas with newspaper.
2. A stick of chalk or charcoal broken in half is easier to use than a long, new stick—and it gives better results. You are not spoiling the chalk or the charcoal by breaking it in half—you are improving it. Half a stick is less likely to break into tiny pieces that have to be thrown away. Also, it doubles your supply.
3. No pencils! Do all sketching directly with charcoal.
4. Show children how to hold the charcoal under their hands for looser, freer sketching. Lay the charcoal flat on the paper—then pick it up (close to the end you are going to use) so the thumb is on one side and the fingers along the other side and slightly below it.
5. Black chalk can be used if charcoal is not available.

6. Sketching implies rapid, loose motions. Encourage the children to lift the charcoal off the paper between each motion; this will help them loosen their hand muscles, and they will be less apt to draw in a rigid and continuous line.

7. After the objects have been sketched lightly, have the children go back over the lines to darken them. Bits of charcoal will break off as they press hard to darken and sharpen the lines, but this is normal usage. Excessive breakage can be avoided by holding the charcoal close to the drawing end.

8. Have children put their names and titles on the backs of the papers before removing them from the newspaper, since the pressure will print the names on the newspaper—or desks, if the newspaper has been removed.

9. If the finished pictures are to be displayed where children's fingers or clothing will rub against them, spray the pictures lightly with fixative. Hair spray makes a satisfactory substitute for fixative.

10. Encourage children to use encyclopedias or other reference books to find additional information about the different forms of sea life. If you have a school librarian, ask her to send a collection of material to your classroom.

Sea Giants

lesson 2
That's Where They Live

OBJECTIVES

1. To be more aware of the variety of sea life.
2. To use cut paper to express an idea.
3. To observe the effect of one color on another.
4. To develop ability to arrange parts into a pleasing whole.

Certainly, everything has a home! That's where they live.

Do any of you have an aquarium at home? Someone surely will, so let him tell about it. Is there one in the school that they have seen—or can they go and look at it?

They have to be little fish, don't they? Goldfish are big compared to tiny guppies, but even the biggest goldfish is small compared to some other kinds of fish. You wouldn't be able to keep a swordfish in your aquarium, would you? There are huge aquariums—bigger than this room —where you could see giant-sized fish. Have any of you ever been to a big aquarium like that? Someone in the class may have visited one. If so, let him tell the class about some of the things he saw.

What kind of aquarium would you like to make? How about one for tiny fish? Well, then you wouldn't need an aquarium any larger than this piece of 12″ × 18″ paper, would you? That's big enough for the little fish you would have at home.

Well, let's see. You are going to have a problem if you want to make a giant aquarium large enough for dolphins and sailfish. Is there any way you could do it with the same size paper—and yet make us know that it is an aquarium larger than this room and big enough for such giant fish? If you make them small enough to fit on this paper, they'll look like little fish—even though they're in the shape of dolphins and sailfish. There must be another way. Now that's a good idea! You could just have the head of the dolphin showing and the rest of him could go right off the edge of the paper; then it would look big. It would look as though you were just seeing a part of the aquarium. If it was a sailfish you'd also have to show part of the fin—part of the sail, so we could recognize him. There might be some seaweed there, too; the aquariums you told about had plants growing in them.

You decide what kinds of fish you want in your aquarium and then arrange them on a paper so they tell us whether it is a big or a little aquarium. Before you begin, though, let's look at something else. When you look into an aquarium you look through glass—and glass is shiny. Our aquariums have to be shiny, too, so they will look like glass.

Have your class gather around you for a moment. Cut off a piece of blue cellophane slightly larger than the 12″ × 18″ white paper and lay the cellophane over the paper. See—water inside shiny glass! Water isn't always blue—so, if you like, you may use green cellophane—or purple cellophane. Each time, cut off the colored cellophane and lay it on the white paper.

There's one more thing you'll have to think about, though. All your paper won't be white because you'll have colored fish and plants on it. The white paper changed color, didn't it—depending upon what color cellophane I put on top of it. Do you think the colored fish and plants will stay the same colors? Do you think they will change to the color of the cellophane? Do you think they will be different colors? Well, let's see who is right.

Tear off several small pieces of different colored construction paper and lay them on the white paper; we'll pretend they're fish and seaweed. Now let's lay the blue cellophane over the whole thing.

There will be exclamations of surprise and delight as they see the new colors appear. Yellow will have turned to green; red has become purple; orange is now gray; blue is a stronger blue. Every color has changed. Talk about how colors affect and change one other. Do the same thing with the green and then the purple cellophane.

Each color makes all the others look different, doesn't it? You will have to decide which color cellophane you want to use, and then you will have to think how it is going to affect the other colors you use.

Give each child a piece of 12″ × 18″ white drawing paper and a pair of scissors while groups of children take turns selecting the first one or two colors they want to use; this will get everyone started in a hurry. When they need additional colors, have them use pieces other children have left over—or, of course, return to the supply area if necessary.

Oh, you don't have to see much of it to know it is a giant fish! What other things are you going to put in with it? Looks like you're making an octopus. Certainly it's all right to have an aquarium of just octopi. Move all your little fish about on your paper until they look as though they are swimming in an aquarium. You have noticed that seaweed bends and moves with the motion of the water, haven't you? Paste everything to the white paper as soon as you have it in exactly the right place; then get rid of all the scraps before you get the cellophane and the glue you need for that. Get a newspaper, too, so any extra glue will get on that instead of your desk.

Remind the children to put a line of glue around the edges of their white papers; then lay the cellophane on top and gently press it in place. The cellophane is bigger than the white paper, but don't cut it off. Leave it there, and when the glue is dry we'll trim off the extra cellophane with the paper cutter.

It will take more time for glue to dry than it does for paste, so this time leave the finished pictures on the children's desks and let groups of children take turns walking about the room to see all of them. Encourage the children to make appreciative and thoughtful comments about each child's work. Occasionally, ask a child to tell what color construction paper was used to create a particular color effect.

There are all kinds of aquariums in your room—tiny ones, and others even bigger than the room. Well, they're for all kinds of fish, and that's where they live!

Make It Easy—For Yourself!

1. Use paste for adhering construction paper parts to the white drawing paper. *Do not* use paste or white glue to attach the cellophane; it will come off as soon as the adhesive dries. Be sure to use a quick-drying cement such as airplane glue or Duco Cement.

2. Have children clear away extra paste and all scraps of construction paper before gluing the cellophane over the pictures; this

will provide more work space as well as make the final cleanup easier. It will help if you carry a 12″ × 18″ paper as a tray while you walk about the room, so children can pile leftover pieces of paper on it.

3. If you have a scrap box of colored construction paper, this will be a good time to use it, or you may want to begin one for use another time. Teach children to save only those pieces of paper large enough to be used another time, and discard the others.

4. Satisfactory paste applicators can be made if you do not have paste brushes. Fold a scrap of paper until it is a narrow strip, strengthening it by bending it in the middle. Do not use paste brushes in glue.

5. The cellophane will tend to shrink slightly as it dries, so don't trim off the excess edges until the glue is thoroughly dry. Trim the pictures with a paper cutter.

6. Begin a collection of pictures of sea life brought in by the children. Encourage older children to find more information about them.

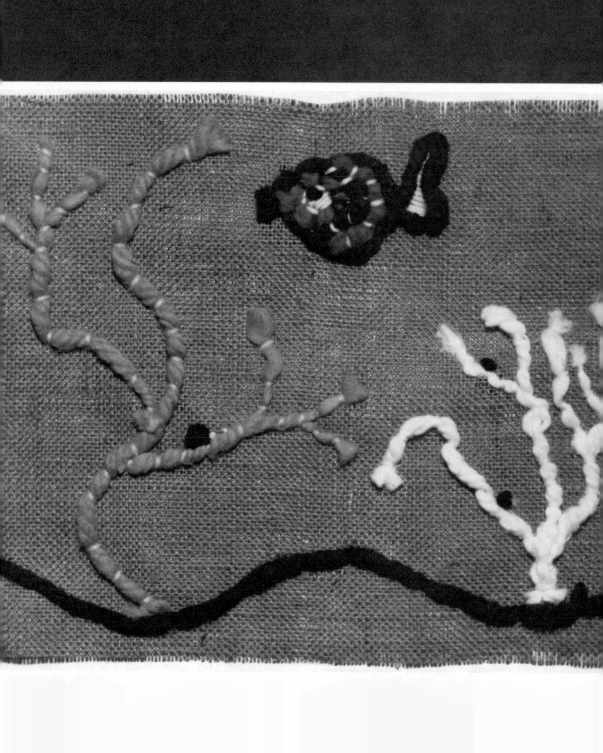

12 Plants

lesson 1
Feel the Sea

OBJECTIVES

1. To develop the ability to use fingerpaint with rhythmic, flowing motions.
2. To experiment with various hand positions and motions to create desired effects.
3. To provide the opportunity for using hands directly in a fluid material.
4. To appreciate the rhythmic feel of lines and motions in an abstract picture.

No, you can't go swimming now. But you will be able to feel the sea—right now!

Whenever you think of things that live in the water, you think of fish, don't you? But are there other kinds of things that live in the water, too—things that stay pretty much in one place. Plants grow in water just as they grow on land.

Talk for a while about plants they have seen in water—seaweed they have found floating in salt water; the plants they have in their aquariums at home; pond lilies; the tiny plant life that grows on the surface of a nearby lake. Does it all look alike? Not at all! The leaves of pond

169

It's Moving!

lilies are large, almost round petals that spread across the surface of the water, but did you ever try to pick one of the lovely flowers of the pond lily? They have long, thick stems that spread out and down into the water, different from the tiny leaves of other plants you see. Have you ever seen watercress growing in a running stream? Sometimes people eat watercress; you might have had it in a salad, or just as a garnish to make other food more appealing. Your mother may have put it in soup, too.

Plants stay in one place, but do they move at all? No, they can't move about by themselves the way fish or animals do, but do they ever move at all? Wind can blow the leaves and stems of plants that grow above ground. That's one way you can make a picture look as though the wind is blowing, isn't it? But what moves plants that grow on or below the water?

Talk about the motion of the water; compare it to the motion of the wind. Is there more motion to water in a pond or in the ocean? Would water in a river always move plants in the same direction? Why is the water in the ocean sometimes rougher than at other times?

Can you use your hand to show the motion of the water in a pond? The water stays almost smooth, so your hand makes almost flat motions. Can you use your hand to make a water lily leaf and stem? Not your finger—your hand. The side of your fist moves in a flat, rounding motion, and then a slightly curving line for the stem would be fine. Let someone suggest motions for the blossom of a water lily.

Ask someone to use his hand to make the motion of water bubbling down a stream. It could flow rather smoothly—or it could bump along more unevenly over rocks. What would plants growing in the brook have to do? They'd bend—perhaps way over if the water were flowing rapidly. The little branches of the plants would all be bending in the same direction, wouldn't they?

By now someone may suggest that the motions they are making are like fingerpainting. If no one sees this, you comment about it. You are really fingerpainting in the air, aren't you?

Can you fingerpaint the motion of water lapping up on the beach on a day when the ocean water is calm? The water would move in gentle waves. Would they all be the same size? No. The side of your hand could fingerpaint that kind of motion. What would the same ocean look like on a rough day? The waves would be bigger and faster, and crash against the shore. Make your hand fingerpaint that kind of motion. Seaweed would be torn loose in those waves, wouldn't it? What would the plants in your aquarium at home look like? No—don't draw them with a finger. Use your hands to show the motion.

See that each child has a shirt-smock on backwards with a button or two fastened in the back; then let the class gather around you while you demonstrate. Cover the desk with a double layer of newspaper.

Let's look at the paper you will use; it is a special kind of paper made just for fingerpainting. Fold a piece of fingerpaint paper back on itself so the class can see both sides of it at the same time. Be sure the children see the shiny side—the side to paint on.

The paper must be thoroughly wet first, so pour a puddle of water in the center of the paper. This is really a handpainting—even though it is called fingerpainting—so lay the palm of your hand flat in the puddle of water and spread it quickly and evenly over all the paper. Look—some of the paper, especially the edges, dried almost immediately, so I'll spread some of this extra water over it again (or you may have to pour another puddle of water). Be sure all the paper is wet; let your hand carry some of the water right off the paper and onto the newspaper. If you have too much water on the paper, just push it off onto the newspaper. That won't do any harm—that's why the newspaper is there.

Use your fingers to scoop out at least a tablespoonful of fingerpaint. You're not going to mind getting your hands into the paint, are you? Of course it feels good—soft and smooth. Place the scoop of paint in the center of your paper, just as you did the water. Lay your hand flat on the paint—but don't spread it yet. Instead, stir it round and round until it feels and looks smooth and slippery; now it is ready to spread over all your paper. Use big motions that carry the paint to and off all the edges—just as you did with the water.

Looks a little bit like the motion of water already, doesn't it? That is because I used motion lines to spread the paint; now let's see what we can do with it. If it were your picture, you'd have to decide whether you wanted to make some plants on top of the water—like pond lilies, or the tiny leaf plants you see on the surface of the water, or seaweed in the ocean, or plants in your aquarium, or grasses growing into the edge of a rushing stream. Suppose it was seaweed—how would you hold your hand? You could make a fist—or you could hold your hand on the side of your little finger. Whichever way you do it, you would move it to make the shape of growing, moving seaweed.

As you talk, glide your hand over the wet paint to create the branches of seaweed. Make all the lines move to one side. Which direction is the water flowing? Oh, look again! It has to be that way because all the motion of the seaweed is in that direction. It is moving quite a bit, so there must be lots of motion to the water. Move your hand in strong, swinging, rhythmic lines. Make another growing piece of seaweed, so the painting has a feeling of both plants and water.

How would the plants in the aquarium look different? They would grow straighter because the water would be more quiet—still. Could you make it look that way?

Erase a section of your painting by sliding the palm of your hand over it. Yes, that would be almost quiet water, so all I need do is grow a plant or two.

Could you make some pond lilies? Erase another section. How would you do that? No, no—no drawing with fingers! Use your hand instead. Yes, you could do it that way. Want to try it here? Would you have just one? You'd want several to make a pleasing arrangement and a painting that looked finished. Would they be just across the top of the paper? No, they could be anywhere on the surface of the paper, just as they could be anywhere on the surface of the water.

How would you make a stream rushing over rocks? Well, no—you wouldn't have to see any rocks. It's the motion of the water and the plants we're interested in. Good! That would be a fine bumpy, moving motion. Would you like to do it on here? You're right! Plants aren't apt to grow in that kind of place, but grasses could grow along the edges of it and bend with the force of the water. Want to try it here?

Well, you certainly couldn't do all of these on one picture, could you? So decide what kind of water you want to show—and what kind of plants might grow in it; then make your picture. You'd better decide what it's going to be before you start to paint, though, because you have to get it all done before any of the paint begins to dry. You can erase your picture once or twice, but don't do it so many times that you rub off that shiny finish. If that happened, you wouldn't have those nice white areas where the paint has been pushed off.

Have the children notice where you have been standing—well back from the work area. Why do you suppose I have been standing so far back? It is important to help me stay clean, but there is another reason, too. Step up close to your painting and show the class that it makes it difficult to move your arm and hand in big, rhythmic motions. Then step back, supporting yourself with your other hand on the desk. See how easily my painting arm can move in any direction!

Are you ready to paint? Have you decided the kind of water and plants—the kind of motions—you are going to make? Fine!

Assign each child to a group with whom he will share water and paint. Make sure all shirts are on backward and buttoned, and that all sleeves are rolled up above the elbows. Give each child the newspaper and fingerpainting paper—and again remind them to stand back from their paintings. You will need to be in all places at the same time for the next few minutes.

Be sure to use the shiny side of your paper. Take turns pouring a little puddle of water on your paper. Oh, that isn't nearly enough water to cover the whole paper; add more right away. Spread it over all the paper. Oops—that's too much water! Push it gently off onto the newspaper. Now a scoopful of paint on your fingers and into the middle of the paper; don't spread it until you have stirred it. Do you need more paint? Spread the paint to the edges of the picture and right off onto the newspaper. No, not your finger! Use your whole hand for painting. That's the way! It looks like rough water. What kind of things will be growing in it? No, you don't need another color for the waterlilies. The lines of painting which your hand makes will show us what they are. Good! It has lots of motion to it! Don't work too long at your painting; stop the second it looks just right.

Have each child put his wet fingerpainting on newspaper that has been placed on the floor. Now for the cleanup: each child should have wiped his hands on a piece of paper towel as soon as his painting was finished (the paper was passed out before the lesson began). Have each child fold his newspaper inward from all directions (with the paper towel inside) to keep all water and paint inside. Let one child collect and empty the cans of water. While these things are being done, you check the jars of paint. If some are empty, have a child discard them. If there is paint remaining in them, combine like colors to make nearly full containers, wipe off the rim of the jar, replace the cover, and have a child put the paint in its proper place. Assign one child in each group to fold the newspaper on the sharing desk and collect all the folded newspaper from his group. Take one last precaution: before anyone sits down, walk about the room with a damp sponge and wipe off any spots of paint that are on chairs or desks. There will be surprisingly little of it.

It may be difficult for the whole class to get around the room to see the wet paintings, but hold up two or three—or point out several in different parts of the room where groups of children can see them. Later, plan a display.

See, you don't have to go swimming to feel the sea! You can fingerpaint.

Make It Easy—For Yourself!

1. Probably the success of no other kind of art lesson is more dependent upon preliminary planning and good organization. Be sure your materials are all ready and easily available before you begin work with the children. Set up rules for procedure so each child knows exactly how to get his materials, what to do with his finished picture, and how to clean up his work area.

2. Cover all work areas with a double layer of newspaper. Two layers absorb extra water and paint better than a single one.

3. Have a sharing desk for each group of four to six children. Two cans of water and a jar of fingerpaint can be shared by the children in the group. Each group of children should stay within their own area.

4. Be sure children paint on the shiny side of the paper.

5. Give each child a piece of paper towel before he begins to paint. He may leave it on the sharing desk or tuck it under the edge of his newspaper, but be sure it is out of the way, yet handy to get to when he needs it.

6. Yellow fingerpaint makes little contrast on white paper—and it is strong contrast which makes an interesting painting. You may choose to avoid using yellow on a one-color painting.

7. It is hard pressure of the hand on the painting that pushes paint off the design and exposes the white paper. Slight pressure during fingerpainting produces uninteresting results.

8. Avoid erasing the design more than once or twice; this tends to wear off the shiny finish and lets the color stick to the paper, producing less color contrast.

9. Always stand to fingerpaint—a step away from the painting.

10. If the painting tends to move while you are working on it, lift one edge and put a tiny bit of water under it; this will hold the paper in place.

11. Any paper that has been wet tends to curl as it dries—particularly the edges. When the paintings are dry, you can turn them upside down and weight the edges or, better still, use a paper cutter to trim off a fraction of an inch of paper all around the painting.

12. If you are working with older children, you may want them to use two colors. Cover part of the paper with one color paint and the rest of the paper with another color. Blend them slightly where the colors meet.

lesson 2
Underwater Garden

OBJECTIVES

1. To be more aware of plant life on the water.
2. To experiment with stitchery as an art form.
3. To use lines to create rhythm and motion.
4. To learn a simple form of appliqué.

Would you like a garden—but not all the bother of planting seeds and pulling weeds? Well, then—how about an underwater garden!

Have you ever seen an underwater garden? The children may think this a strange idea, but assure them that there are such things. And, yes, I'm sure you have seen some. Some child may have taken a trip in a glass-bottomed boat—or at least heard about them. What do you think you would see if you could look far down into the water? There would be fish there—all kinds of fish, but there would be other things, too. If the water wasn't too deep you would be able to see the bottom, and there might be sand and rocks; there might be shells and even coral.

Things grow in a garden, so if this is an underwater garden something should be growing in it, too. You would see plants—seaweed—growing. Would it grow straight up the way plants in your backyard garden do? Well, probably not. The water would be moving, so the plants would have to bend with the direction of the water, the way the wind sometimes blows grass, or wheat, or flowers.

A Garden without Weeds

I said all of you have seen underwater gardens—and all of you haven't taken trips in glass-bottomed boats, so where else might you have seen some of these things? If you have an aquarium at home—or have seen one somewhere else—you probably have seen an underwater garden. Do you have plants in your aquarium? And sand or stones or shells? Then you have seen an underwater garden. Perhaps some of the children have visited a giant aquarium; let them tell about it. They will be especially interested in the fish that were in it, but bring their attention back to the plant life that was in it, too. After all, the plants make the garden for the fish to live in, don't they?

Are there other kinds of underwater gardens that you may have seen? Talk about going swimming in the ocean and the seaweed they may have seen, or the plant life in ponds and lakes. Sometimes you can even see plants on the surface of the water. You can sometimes see lovely pond lilies, but if you ever tried to pick them, you know that the stems go deep down under the water. There are small plants that drift near the surface of the water—or even on top of it—that are called plankton; perhaps you have seen that.

Explain that they are going to make an underwater garden picture. Show them the cotton roving they will use. Could you make some underwater or on-top-of-the-water plants with this? See how easy it would be to make seaweed with this and to make it move with the motion of the water. Cut off a piece of cotton roving and make branches that can be moved as you talk. Could you make the shapes of leaves and flowers?

You're not going to use paste to stick them to paper. In fact, you're not going to use paper, either—you're going to use cloth for the background. Show them the several varieties of colored burlap or other heavy material you have. You won't glue the lines to the cloth. There's only one other way to get them to stay where you want, so you will sew them in place.

You won't be able to sew this heavy cotton roving through the cloth, so you will sew it to the cloth—you will appliqué it. Let's see how we do that.

Let your class gather around you. I'll use this dark-colored roving on light-colored burlap so the lines will show well. Cut off a piece of roving long enough to make an important part of the picture, but not so long that it will be difficult to handle. Lay it on the cloth and move it about to make the beginning of a water plant. Thread a large eye tapestry needle with a piece of ordinary yarn the same color as the cotton roving. It will need a knot in one end of it.

Pull the needle through from the back of the cloth so the knot is underneath the cloth and the yarn is close to the side of the cotton roving. Put the needle over the roving and down through the cloth. Pull the yarn through the cloth until it has made a stitch over the roving. See— it has held it to the cloth.

Let's try that again. Move the needle a short distance—not more than an inch—and do the same thing again. See—up on one side of the roving— over the top—down on the other side. Move the needle—up—over—and down again. See what is happening? The big, fat yarn is being held to the cloth. Bend the roving in another direction and take another stitch. See how easy it is to draw with the cotton roving and fasten it to the cloth with yarn!

Have you noticed that the stitches hardly show at all? Why is that? I used the same color for both the stitch and the line, so there is no contrast. Could I make the stitches show if I wanted them to be an important part of the picture? Of course—just appliqué the roving with yarn of a different color. Thread a needle with a contrasting color and make several more stitches over the roving. See how different that looks! When you make your underwater garden you will have to decide whether or not you want the stitching to show. Perhaps you will want it to be seen on some part of the picture and not seen on some other part.

This isn't the kind of picture you can plan ahead of time; you will have to make it grow as you go along. Well, that's the way things really do grow, isn't it?

Leave the cloth, cotton roving, and yarn where the children can see the variety of colors and choose from them. Let them take turns selecting their beginning materials while you give each child a large eye tapestry needle. Remind the class to make the largest thing first; that will help organize the space on the picture.

Just decide what you want to begin with and where it will start. It is like drawing a line with crayon; start the line any place you want, and then add to it. That's right! Up through the cloth, over the line, and down the other side. You can make some of the water first if you want to. Good! That is a fine beginning for a water lily. Could you make your stitches closer to the cotton roving? It will make the picture look better and hold it more firmly in place. That's much better! That's a fine idea! The water under the line of the plant is in back, and the one on top looks like water in front of it. It takes longer to make pictures that are appliquéd than ones that are just pasted, but don't they look nice?

It will probably take more than one art lesson to finish the pictures, but that is no problem. Just stick the needle through the cloth and leave it until next time.

When the pictures are finished, staple them to a construction paper background and display them. Isn't it nice to enjoy a garden without all the fuss of planting, pulling weeds—and getting rid of bugs? An underwater garden is just the thing!

Make It Easy—For Yourself!

1. Any kind of yarn can be used for the picture—cotton roving, bulky yarn, or even regular weight yarn. For upper grade children, you may want a variety of kinds to add interest and contrast to the pictures.
2. Cut the burlap or other heavy background cloth to pieces 10″ × 12″ or larger.
3. There are giant tapestry needles about three inches long that have extra large eyes. These are especially good for young children to use.
4. Very young children may need help threading needles unless they have the extra large size.
5. Demonstrate how to thread a needle with yarn—and let each child try it. Stretch the yarn over the side of the needle, leaving one end hanging down five or six inches. Use the thumb and forefinger of your other hand to squeeze the yarn and needle together. Pull the needle out so you are tightly holding the folded piece of yarn; this way there are no fuzzy edges on the yarn. Press the eye of the needle between your thumb and finger so the thread is pushed through it. Pull it through until it is a single thread. Don't put the yarn in your mouth to get it wet as you might do with thread.
6. Leave several pairs of scissors at the supply area so children will not have to carry theirs with them.
7. Frequently encourage each child to hold his picture as far away from him as possible in order to get a better view of his work.
8. It will probably take more than one art lesson to complete the picture. Save it for another lesson or let children work on their own during extra time. Do not let the children take them home to work on them, since you want to see their progress and assist as needed. You may want to have a special place where needles can be kept so they won't be lost between work periods.

13
Oceans, Rivers and Lakes

lesson 1
Drifting Along

OBJECTIVES

1. To be more aware of the variety of things found in bodies of water.
2. To provide opportunity for contributing to a cooperative project.
3. To improve ability to create with cut paper.

It's a lazy, sunny day. You're in the mood for quietly drifting along on the water. You'd enjoy that.

Let's pretend we've been somewhere else today. Some of you were taking a boat ride down a river; some of you were jumping over rocks in a brook; some of you were swimming in a lake. You were all having fun around water. There were all kinds of things to see in the water; what were some of them?

If you were lucky, there were fish and you might even have caught some. What are some other things? Let's see if everyone can think of one thing that is different than that which anyone else saw—something you can show later in a picture.

Ask all the children to stand. Go around the room letting one child after another name something he might have seen in the water that he would like to make. As they name one different thing, have them sit down. In this way, you will know who still doesn't have an idea—those who are still standing. Don't take too long; if a child doesn't have an idea, move quickly to the next child. Later, come back to the children you skipped.

Such things as turtles, frogs, and snakes will probably be mentioned first. When there is some hesitation, ask a question. Are there any birds that belong around a lake or pond? Certainly—ducks, swans, and gulls. Would there be any kind of plant life in the water? There would be weeds, grasses, and pond lilies. Would there be any water animals you might see if you were lucky?

Each question will direct the children's thinking, and from there lead to other ideas. If you went swimming you would be in the water, too! If you were in a canoe or a boat, that would be there.

Try to have each child express an idea that is different from everyone else's. If necessary, let two children make the same kind of thing, but remind them that they won't look just alike.

Show the class the assortment of colored construction paper that is available. There will be questions, so take time to answer them. Yes, you may make them any size you like. It won't do any harm if your waterbug turns out to be larger than his beaver; just make them large enough so we can see what they are, but not so large that they are hard to handle. No, they don't have to be the color that real salamanders or snails are. You are not making real salamanders or gulls or rocks. You are making pictures of them, so you may make them any color you like. Of course— if you like, you may make them realistic colors. When they are finished we'll put them all together; perhaps you will be able to think of a good title for them.

Let groups of children take turns selecting their beginning papers; one or two colors will be enough to get everyone started in a hurry. While this is being done, give each person a pair of scissors so he can go to work immediately. You can give pasting supplies to them as you walk about the room to help individual children.

Oh, but that is such a tiny pond lily, it is hard to be sure what it is! Could you make a larger one so we can see how lovely it is? I can tell right away that it is going to be a beaver! His flat, broad tail tells me so. Rocks are easy to make, aren't they? Would you like to make another, larger one? You may make something different if you like. Um-m—crayfish are something like lobsters, aren't they? Could you give him giant front claws? That's the way to do it! It wouldn't be a good thing to meet that alligator, would it? Don't you need a different color for his eye? Gray on gray doesn't show. Right! That's because they are both the same, so there isn't any contrast. Good! That shows much better; just paste it over the other one. Paste the parts together as soon as they are ready.

Continue to encourage and help children as they work. Maybe you can even think of something no one else is making. Yes, when you have finished the first thing, you may make another thing if you like. It is more important, though, to do a good job than to make several things.

It was fun to make all the things they might find in streams or ponds or lakes or rivers. It will be fun to show them—and to tell about them, too. There won't be much for any one child to say, but encourage him to add one detail about his object. As children talk, jot down the things they say. Later use the things that were said when you display their work, or use them as the basis for a language lesson. Make it a class project to group similar ideas together; to select the more interesting things that have been said; to add a new thought. Show the class how to group phrases or sentences into poetic form; then use the class poem for a bulletin board display and group the pictures informally on or around the verse.

You will be as pleased as the children are; besides, you can pretend it's a lazy, sunny day any time. Just lean back and let yourself drift along.

Water is a wonderful place
To see all kinds of things—
A crayfish with giant claws
That won't let his food get away.
Pond lilies float
On green lily pads;
Plants sway
As fish glide silently by.
A snail creeps along
Even slower than the turtle.
Alligators are fine
To see from boats—
But don't go swimming with them!

Wouldn't you like
To be a swan?
Would you dive under the water—
Or would you flap your wings?
Or would you rather be
A green salamander—
Or a jumping frog—
Or a rock
That had nothing to do?

Make It Easy—For Yourself!

1. No pencils! No preliminary drawing. Don't let children add details with pencil or crayon. If the part is too small to be cut from paper, it is too small to be seen.
2. Encourage children to share paper other children have left over when they need only small pieces. Of course, let them return to the supply area whenever necessary.
3. Teach children to do their pasting on newspaper, in order to keep their pictures and work areas clean. In this case, only a small piece of newspaper is needed. A large paper to cover the whole desk would be a nuisance, and it could easily pull all the work off the desk.
4. If paste brushes are not available, show the children how to make paste applicators by folding scraps of paper until they are narrow strips, bending them in the middle to give added strength.
5. Some children who make simple things will have time to make a second object. Let them make any found-in-the-water things they would like. Encourage children, however, to do their best work rather than hurry to make more things.
6. Display all pictures with the verse or some comments about them.

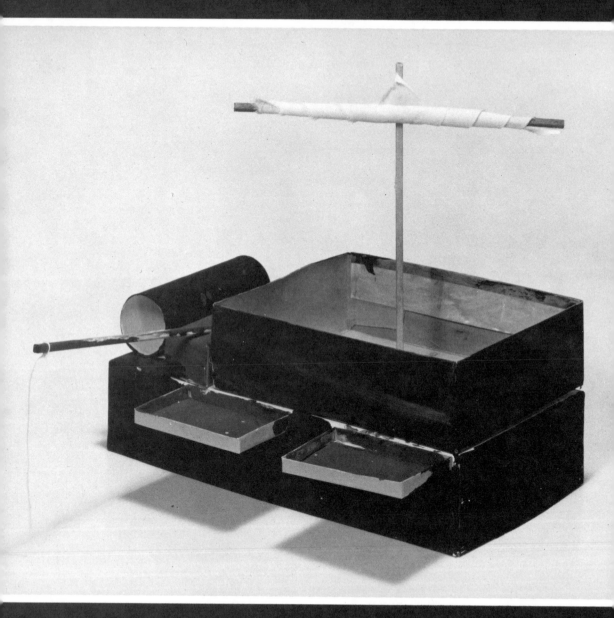

Whaler

lesson 2
At the Marina

OBJECTIVES

1. To encourage children to use found materials in their art work.
2. To provide opportunity for three-dimensional construction.
3. To experiment with a simple form to create a specific kind of boat.

You can't believe your eyes! You never knew there were so many kinds of boats. Look at them all tied up at the marina!

Have you ever been to the docks—or to a marina—and seen all the boats that come and go? It is a busy place, isn't it? What kinds of boats did you see there?

Let the children talk for a few minutes about boats they have seen. They'll want to tell about boats they have been on, too. Encourage them to add some detail to help other children visualize the boats they describe. Have you ever seen an ocean liner? It would be very different from a yacht, wouldn't it? Oh, yes, much much bigger! Right—a yacht wouldn't have huge smokestacks, either. How is a motorboat different from a rowboat? It might be no different, except one has a motor and the other has oars. Some motorboats are larger and the sides curve more.

Hold up a small, shallow box, about the size candy bars come in. This doesn't look much like a boat, does it? Could you change it, though, so it makes one of the boats you have been talking about? A mast and sail would help it look like a sailboat; you could even add two masts and make it a schooner.

Have some small dowels or applicator sticks and cloth that you show the children; they would make good masts and sails. Could you use any of these smaller boxes—or these cardboard tubes in any way? That's a good idea! That tube could be attached to make a paddle wheel boat! Could you make an upper deck, too?

Continue to show various size and shape boxes as well as the dowels, cloth, and construction paper that is available, and encourage children to think of ways of combining them to create specific kinds of boats. Explain that they will use glue to attach the various things, so they will have to be patient and hold the parts together until the glue dries. When you have finished building your boat you may paint it—every well-made boat has to be painted!

Ask each child to think of the kind of boat he would like to make. Let groups of them take turns selecting the boxes and other materials they will need. In the meantime, give each child a tube of glue, a half page of newspaper to work on (to keep glue off his desk), and a pair of scissors.

Certainly, it's all right to make a rowboat, if you like. You will have to make some oars for it. How will you make seats in the rowboat? Good! Those fat cardboard tubes make excellent smokestacks. How did you get them to slant so nicely? I see! You cut off a slanting piece at the bottom. Sails aren't always up, are they? Certainly not when the boat is in the marina. Rolling the sails was a good idea. Do you need some help? I'll hold that for you while you add the other piece. You have to hold it steady so it will dry strong. No, most boats aren't straight across the front and back like these boxes are, but you can still tell what kind of boats they are. Anyone can see that this is a cabin cruiser. You can add a pointed bow and stern if you like.

Hold up a boat occasionally and comment about some part of it that is unusual. It may be the combination of materials that is unique—or the special way one of them has been used. There may be a particular detail that is especially good, or a child may have been successful in solving some problem.

Continue to help each child in whatever way he needs—a question or a suggestion, a bit of encouragement or a compliment, or just a third hand when needed. Have a painting area ready for the children to use when they have finished constructing their boats. Children should share painting supplies and then take their boats back to their own desks to dry—or to a special area you have covered with newspaper.

Arrange an exhibit space large enough to display all the boats. If there is not enough space in your classroom, look around the school for other suitable places. Is there a display case in the school lobby? That would be an ideal location. The librarian will be able to plan space for you—and perhaps add a collection of books about boats. The principal will welcome some for his office. The school nurse or social worker will be happy to have children's work displayed.

There are lots of places in the school for the boats. After all, who wouldn't like to have his own boat at his own private marina!

Make It Easy—For Yourself!

1. Take plenty of time for motivation. It takes lots of ideas to transform an ordinary box into a special kind of boat.
2. Have a large assortment of small boxes. A local store where candy is sold can supply all you need. Save cardboard tubes from such items as paper towels, toilet tissue, waxed paper, masking tape, and mailing tubes, until you have a good variety of sizes. Use construction paper and scraps of cloth of any color. Applicator sticks, thin dowels, or thin strips of wood make fine masts or other parts. Other scraps of wood, even large buttons, may be useful.
3. Use glue to attach the parts. Paste will not hold wood or buttons and is not strong enough for cardboard.
4. You may want to plan two art lessons to complete the project. See that each child gets a good start before ending the first lesson. Complete the boats and paint them during the second lesson.
5. Prepare one or more painting areas where children may share supplies, rather than have each child work at his own desk. One long table will be enough—or push several desks together to make a number of smaller painting areas. Cover them with newspaper. Have several cans of water for washing brushes, and several palettes.
6. Egg cartons broken in half make fine palettes.
7. Use large wash brushes for painting the boats.
8. Don't worry if some of the lettering from the boxes shows through the paint. The clever constructions are the important part.

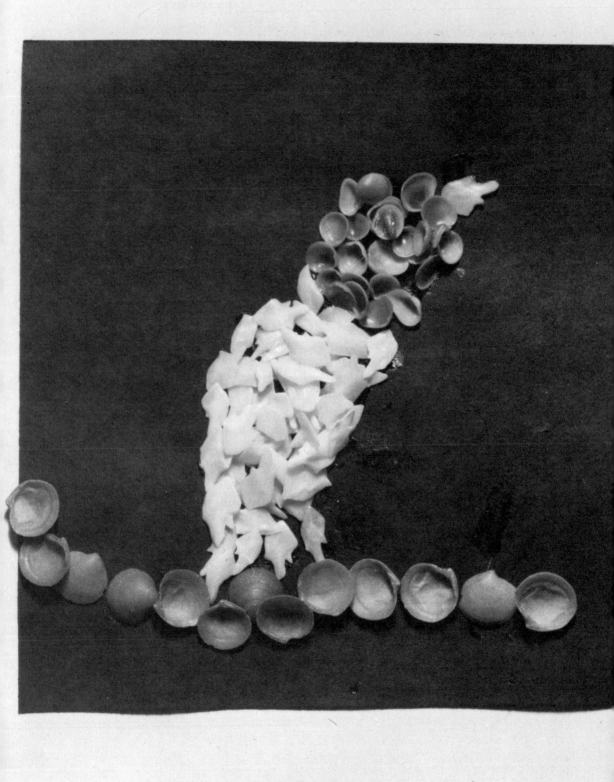

14 Shells
and Sand

lesson 1
New Fossils

OBJECTIVES

1. To use natural materials as part of an art lesson.
2. To experiment with an unusual combination of art materials.
3. To create a three-dimensional art object.
4. To introduce a simple casting technique.

It takes hundreds of thousands of years for fossils to be created. They're a rarity to find. Well, most of the time that's true—except new fossils.

Do you know what a fossil is? You may have to explain it to younger children—better still, show them one if you can. Older children will know what a fossil is—and someone may even have one at home. Perhaps he will be able to bring it to school to show to the rest of the class. If there is a natural museum nearby, there will probably be several fossils there; some of the children may have seen them. Encourage the other children to go see them, too.

Let children tell what they know about fossils. Be sure to bring out the information that fossils may have the object itself imbedded in the rock, or they may have the imprint of the object left in the rock.

Let's make our own fossils. We won't be able to make the stone, but we can use plaster of Paris instead. We'll use shells for the fossils, and you

197

All That's Left

can make the kind of fossils that will have the shell left in them—or you can make the kind that will have the imprint of the shell in the fossil. Perhaps the shells you use will help you decide which kind of fossil you will make.

Have your class gather around a table so you can demonstrate. You will need some nonhardening clay for the beginning. Soften it by squeezing it between your hands until it is soft and pliable. Lay it on the table and gradually flatten it, while you bend the edges upward to form a bowl-like container about the size of half a large orange.

That is just to hold the plaster of Paris until it dries. Now let's get to the shells that will be a part of the fossil. Pick out two or three shells— or parts of shells—that are small enough to lay in the bottom of the clay container.

Should I lay the top of the shell against the clay—or should I put the inside of the shell against the clay? Well, let's see about that. Remember that wet plaster of Paris is going to be poured into the clay container—on top of the shells. It will fill in all around the shells, won't it? If the inside of the shell is down—pressed into the clay so no plaster can get under it— we will see the inside of the shell when the clay is removed from our fossil. If you want to see the outside of the shell, which side will you press into the clay? Right! You'd press the outside into the clay. Suppose you have a round shell—such as a snail shell? You can lay it down any way you want; just be sure it is pressed far enough into the clay so no plaster of Paris will get under it, since that would just hide the shell.

Is there any way you can make the kind of fossil which is just an imprint—where the shell itself wouldn't be a part of the fossil? No, you won't be able to just pull the shell out after the plaster is dry; that would break the fossil. If you want to take the shell out, you will have to do something to it first. What does your mother do to a pan before she pours a cake into it—so the two will come apart later without breaking the cake? Right! She greases the pan. If you want to take out the shell, you will have to grease it.

Can you think of any other way you could make a shell imprint in the plaster? Someone may suggest the possibility of pressing the shell into the clay—then carefully removing it. Of course, the shape of the shell would be in the clay, so the plaster would fill that shape. When the clay and plaster were separated, there would be the imprint of the shell on the plaster.

Someone may ask if you have to grease the shell in order to remove it from the clay. That is a good question. No, you won't have to do that, because there is grease in the clay already. It is an oil-base clay—which is the reason it doesn't get hard and can be used over and over again. That is also why it will separate from the plaster of Paris.

Arrange the shells in any way you like, as long as there is no space for the plaster to get under any shell.

Now for the plaster of Paris. Use a large paper or styrofoam cup, about half full of water. Gradually add plaster of Paris while you stir it with a tongue depressor or other flat stick. It will probably take more plaster than you expect, but continue to add it gradually until the mixture begins to *slightly* thicken—to the consistency of thin mayonnaise or thick cream soup. You will probably notice that it has become slightly warm, a chemical reaction that takes place. Show the children what it looks like at that point. Immediately pour the liquid plaster of Paris into the clay mold.

Take time to answer any questions. No, you won't be able to take it out of the clay container for quite a while—several days at least. It will seem firm in just a short time, but it will break easily until it is thoroughly dry and hard. Certainly it could break even after it is completely dry, but normal handling won't harm it then. Yes, the shells will stay in the plaster—except the ones you grease. They will come out of it like a cake comes out of a greased pan. Yes, the clay can be used over again for something else. A very large shell? Yes, the best plan would be to put a long strip of clay under the edge of the shell—the inside of the shell facing down. Press the shell into the clay, and bend the outside edge of the clay up to form a collar around the shell. Of course, the shell would be greased so it could be removed from the dry plaster mold. There would be a big, hollow shape in the plaster mold, the same shape and texture as the outside of the shell.

There is one more thing to know—how you are to clean up after you have poured your plaster of Paris. The most important thing is *not to let one little bit of plaster get into the sink*. The containers and tongue depressors you use for mixing the plaster will be thrown away—so they won't get any plaster in the sink. You will dip your hands into a pail of water if you have any plaster on them, then wipe them on a paper towel and put the towel in the wastebasket. The pail of water will be emptied outdoors, where it won't do any harm. So, no plaster from anything will get into the sink.

Everyone will be eager to begin making his shell fossil, so give each child a lump of clay and let him begin softening it. You have to get it soft before you can begin shaping it into a container. Don't pound it; that doesn't get it soft. Squeeze it back and forth between your hands until you can feel that it is easy to change the shape. You had better choose your shells first so you will know what kind of a container to make. How many shells? Doesn't that depend upon the size of them? One big shell would be enough—or several tiny ones. You will have to judge that by the amount of clay you have and the size of the shells you choose.

Be sure that each child has a container deep enough to more than cover the shells he imbeds in it. Be certain, too, that the shells are pushed far enough into the clay so no plaster of Paris will be able to get under them.

As each child completes his mold, let him cover his desk with a page of newspaper and then go to the mixing area to make his plaster of Paris mixture. Supervise the area carefully to see that materials are handled correctly and the plaster is mixed to the proper consistency—and then poured. If this is the children's first experience with plaster of Paris, they will feel insecure and want your constant assistance and approval. When the plaster is the right consistency, have the child return with it to his own desk—and pour it into his mold immediately. Don't move it unless absolutely necessary until the plaster has set. If the molds must be moved, slide them onto a piece of wood or heavy cardboard so the plaster will not be disturbed.

Wait several days or a week before removing the plaster casting from the mold. Be sure it feels dry and hard. Gently peel off the clay—and save the clay for another lesson. You may want to add a felt backing to the flat side of the plaster so it can be used as a paperweight or table decoration.

Yes, there are such things as new fossils. Here's the proof! It may take thousands of years to make most fossils, but you know a shorter way!

Make It Easy—For Yourself!

1. It is essential that you have all your materials—and plans—organized before you begin the lesson: nonhardening clay; shells (tiny to medium size are the best); newspaper; plaster of Paris; paper cups for mixing plaster; other paper cups for dipping out water and plaster; tongue depressors; pails of water (for mixing plaster and later to rinse hands); paper towels.
2. Explain the whole process in the beginning so each child will know why he is to do—or not do—certain things.
3. A few drops of vinegar added to the water will slow the setting process. This can be an advantage when you are working with many children at one time. More water will not thin the plaster once it has set. If it hardens before it can be used, throw it away and start again.
4. Don't let any plaster get into the sink; plaster will harden even in a pool of water. It is a good plan to cover the sink with newspaper as a reminder not to wash hands or containers there.

5. Don't worry if a layer of water forms on top of your mold. There just wasn't enough plaster of Paris in the mixture, so the extra water has come to the surface. When the plaster is firm, tip the mold and drain off the excess water—outside, of course.

6. If you like, you can add a tiny bit of tempera paint to the water before adding the plaster to tint the mixture. Don't overdo it.

7. Be sure children have the top edges of the clay mold reasonably even—at least all edges well above the shells so enough plaster mixture can be added without draining off.

8. You may want to use two lessons to complete the project if you are working with young children: (1) make the clay mold and place the shells, (2) be sure shells are securely in place, mix and pour plaster.

9. Arrange a mixing area where children can mix their plaster of Paris. Have a large table covered with newspaper. Several paper cups should be available to scoop out dry plaster only, and several other paper cups should be available for dipping out water only. Don't use the same cups for both wet and dry materials. Have a supply of tongue depressors or other flat sticks. There will be less confusion as well as less chance of accidents if all the mixing is done at one large location; also, it will make it possible for you to do a better job of supervision.

10. Don't do any cleanup until the last mixing has been done. Have one child walk around the room with his cup of extra plaster. He should collect all the tongue depressors and put each one in his cup so no plaster drips off them. Then go around the room, stacking cups with any leftover plaster still in them. Occasionally wrap the stack of cups and put them in the wastebasket. Ask two children to clean the mixing area: (1) put away extra plaster, cups, tongue depressors; (2) empty the pails of water in the schoolyard where they won't cause any problems; (3) stack and fold the newspapers with any spilled plaster inside them.

11. Encourage children to seek other information about fossils. Ask the school librarian for help. Begin a collection of fossils or pictures of them.

At the Sea with Brother Crab

Mixed Media Pictures

lesson 2
A Lonely Beach

OBJECTIVES

1. To create the mood of a place in visual form.
2. To experiment with an unusual combination of materials.
3. To encourage children to use natural materials in their art.

There is nothing more deserted than a lonely beach—sand and water as far as you can see.

Remember last summer when the beach was so crowded you could hardly find a place to stretch out on the sand? It seemed like the busiest place in the world, didn't it? What do you suppose it would look like now?

Yes, it would be deserted, wouldn't it? Pretend you go back there now. Would there be anyone else around? No—as far as you could see, there wouldn't be another person in sight. You'd have the whole place to yourself! Do you think it might look bigger than it did last summer? It would look like it went on forever—just sand and water and sky. Would you like it?

Let children talk about their feelings when they are alone. Some would not like it; others would enjoy it. Ask them what they would do at the beach if they were alone there—and it was too cold to go swimming. It might be pleasant just to walk close to the water and listen to it as the waves lap against the shore. Would there be any other sounds? There would probably be seagulls and you would hear their harsh calls. That would make it seem even lonelier, wouldn't it? Would there be anything else on the beach? Encourage children to think of other things: a sandpiper wading along the edge of the water as he looked for food; a crab moving awkwardly over the sand; driftwood; rocks and shells showing the crest of high tide; your own footprints left in the sand. Visualize the vastness of space and the smallness of things. Create a mood of loneliness and emptiness.

How would you make a picture of a deserted beach? Of course, just put in sky and water and sand—and maybe some footprints and a shell, or a seagull, or one of the other things. You wouldn't put many things in, though, would you?

Sky would be at the top of your paper, but would the sand or the water be at the bottom of it? The sand could be there, but would it have to be? Yes, if you were standing on the beach you would see sand—then water—then sky, so that is the way it would be in your picture—sand at the bottom, sky at the top, and water between the two of them. You could be all alone in a boat on the water; then you would see water at the bottom of your picture, sky at the top of it, and the sand would be in the middle.

Talk about the space for each of the three parts of their pictures. Would it be interesting to have all of them the same size? No, no—not at all! When you are at the beach you see lots of sky, don't you? So you may want that to use lots of the space on your picture. Then decide whether you want to see more sand or more water; that way, all three parts of your picture will be different sizes.

Explain that the sky and the water will be painted—but the sand will be made with real sand; it will be a real beach. Talk about the color of the sky. On a clear day when the sun is shining the sky is blue, but is it as bright a blue as your paint? No, it is a light blue. How can you make it a lovely, clear light blue? If the children are to use tempera paints, they can add white paint to the blue. If they are going to use watercolors, they will add water to the blue paint to thin it. It could be a stormy sky; in that case you would have to add gray to the blue. How would you make gray?

What color is the ocean? Are you sure it is just blue? Right! You would need to add some green paint—or perhaps purple, or even black. That would make it look much more like water.

Give each child a piece of 12″ × 18″ white drawing paper and painting supplies. Remind him to plan the place for each of the three parts of his picture. Which part will be the biggest? Which part will be the smallest?

Yes, it is a good idea to wet the sky before painting it. That will help the color blend and look lighter and softer. Paint rapidly so the sky will stay wet while you blend the black into it. It looks as if a storm is coming up, doesn't it? Remember—the sky looks lighter off in the distance, so just add water to that part to thin the blue paint that is on your brush. If the sand comes between the sky and the water, you will be able to paint the water right away, but you will have to wait to paint the water if it is next to the sky. Why is that? Because wet colors painted against other wet colors run together. You don't want the water running up into the sky, do you! Wouldn't the sky be straight across where it appears to meet the water? That is called the horizon. When you paint the water you can cover up that uneven bottom sky line. The water will be darker than the sky, so it won't show where you paint over that bit of sky. Good! You are blending those colors nicely for the water. It looks like real water, doesn't it?

As soon as each child has finished painting the sky and water parts of his picture, clear away all the painting supplies; this will give the paint a chance to dry. In the meantime, cover a table with newspaper, and have one or two boxes of sand on it. Have your class gather around while you demonstrate. Dip a large wash brush into water and then into a container of polymer. Hold the container of polymer over your picture so no drips will get on the floor. Immediately paint with the polymer all the area that is to be covered with sand; put on enough polymer so the whole area will be thoroughly wet.

Explain that this part of the work will be done at each child's desk—and that several of them will share one brush and one container of polymer. As soon as the polymer has been applied, the child will bring his picture to the sand area. Lay your picture into the box of sand. Scoop up some sand on a piece of cardboard and drop it over the wet polymer. It won't do any harm if some of the sand gets on the sky or water if the paint is dry. Lift the picture and drop the surplus sand back into the box. See—there it is—a real sandy beach!

Children will be excited by the technique and delighted with the results—and eager to try it on their own pictures. Distribute the supplies and carefully supervise the sanding area. Be certain that all of the area to be sanded is thoroughly wet. If parts of it have begun to dry, have the child repaint the entire area with polymer. Let each child use the cardboard to scoop up some sand and drop it on the wet polymer. He should immediately drop off the extra sand and return to his desk with his picture.

Now that you have the three parts of your picture—sky, water, and beach—are they all finished? No. We talked about other things you would see—things that would make the deserted beach seem even lonelier. Remember what they might be? Well, you need to put at least one in your picture. Two of them would be fine, but don't put in so many that it doesn't look deserted.

If you are going to make footprints in the sand, do it right away before the polymer dries—just use your fingernail to move small areas of sand. They might be your footprints in the sand—or they might be prints made by the gulls or sandpipers. If you want a shell or rock, get one and glue it onto the sand. Yes, there are some real ones here. They have to be tiny ones because you have a tiny beach on your picture. Besides, if they were big they would be too heavy for the paper.

Have some tiny pieces of white and black paper for the children who want a seagull or a sandpiper or a crab for their pictures. Make them tiny, too, and find just the right place for them. If there are gulls in the sky, paste them in place; but if it is a crab or a sandpiper on the sand, you will have to use glue to attach it. Remember, though, one or two things will make the beach lonelier than many things.

Before long, all the pictures will be finished. Give them some extra time for the polymer and glue to dry while groups of children take turns walking around the room to see each one. Call the class's attention to an especially good blending of colors for the water, to an interesting sky, to an unusual detail on the beach. Later, when they are thoroughly dry, you will want to display them. The sand won't fall off the picture.

Sand, water, sky—a tiny crab, a soaring seagull. Nothing is lonelier than a deserted beach.

Make It Easy—For Yourself!

1. Let young children use tempera paint and older children use watercolors.
2. You may prefer to take two lessons to complete the picture. Discuss the picture and paint the sky and water during the first lesson. Apply the sand and add the lonely details during the second lesson.
3. Cover desks with newspaper to protect them from both the paint and the polymer.
4. Younger children should use easel brushes and older children should use large wash brushes for both the painting and for applying the polymer.

5. Have four or five children share a container of polymer and a brush for applying it. This will cut down on the number of brushes needed—and to wash later—as well as limiting the number of children at the sand area at any one time. Otherwise, the polymer would dry while they were waiting their turns to add the sand.

6. Brushes should always be put in water *before* putting them into polymer, and they should always be put back into the water whenever they are not being used in polymer. Leave them in the water until they can be thoroughly washed to remove every trace of polymer; this means they will need to be washed several times. Polymer is an adhesive, and if any of it is left in the bristles, the brush will be ruined.

7. A solution of half white glue and half water may be substituted for polymer. Apply it the same way—and wash the brushes thoroughly.

part four
The Universe

15 Earth

They're Here, Too!

lesson 1
We Live Here

OBJECTIVES

1. To concentrate attention on the variety of types of things on earth.
2. To experiment with a combination of textured materials.
3. To be aware of the possibility of using found pictures with their art.
4. To develop ability to arrange parts into a pleasing whole.

Look around you! Everywhere you see cars, trains, busses, trucks; houses, stores, factories, offices; birds, animals, trees, flowers. But don't forget—we live here, too!

If you made a list of all the things found on this earth, how long a list would it be? A hundred things! Is that all? A thousand? A million? It would go on and on for a long time, wouldn't it! I'm sure it is impossible to think of all of them, but let's see if we can make a list with as many things on it as there are people in this room. Do you think you can do that? Of course, you can!

Ask every child to stand. Explain that as soon as he has added something new to the list he may sit down. You'll have to be fast, though! Begin at one side of the room and go quickly from one child to another.

Oh, we'll have to skip you if you are slow—but we'll come back again, so be thinking of several things. People, houses, cats, dogs, elephants, stoves, chairs, cups, trees, flowers, toads, bees, ants, rocks, sand, mountains, rivers —it will be easy to name things. Oh, there are many more—and we could go around the room several times before you ran out of ideas. But that's enough to remind us that there are many, many things on this earth.

Explain that each child will make a collage that shows there are many kinds of things on earth. You won't be able to put as many in your picture as we named; yes, I'm sure you could make them all, but it wouldn't look right, would it? You choose several things you would like to have in your picture; then you will cut them out of construction paper —or one of these other materials. Show them the assorted supplies you have: sandpaper, cork, corrugated paper, cloth. Two or three other materials, in addition to several colors of construction paper, will be enough.

There is one other way of getting a picture for your collage. Show them the magazines and flip through the pages of one of them. Look! There's a lady and a little boy—there's a house and a refrigerator—and there's a motorcycle. Look—there are pictures of all kinds of things you find on earth. You may take a magazine and find one or two things that you would like to include in your collage.

Explain that most of the things in the collage will be made from the various materials that are there, but one or two pictures from a magazine will be included. Make several things that you want to be a part of your picture, then find a magazine picture or two and arrange them so they look right. You may find that you need one or two other things, so you will make them, too. It will be a good idea to have some of the things overlapped; it will help make them look as though they belong together. You can also help make things look like they belong together by placing them close together.

Let groups of children take turns selecting their background color and two or three beginning materials. You may take a magazine now, if you like, or you may wait until later to get it. The only other material anyone will need immediately is a pair of scissors. As you walk about the room to help children with their work, you can give them the pasting supplies. Remind them, however, not to paste anything to the background until they have a good arrangement.

Don't worry about all the things you are going to make. Start with an easy thing to cut for a beginning. Good! That's a nice big tree—and trees are important things on this earth, aren't they? Oh, that is a wonderful football player you found in the magazine! You like to play football, don't you? Cut out around him carefully. Wouldn't it be a good thing to start to arrange some of your things? Then you will need to know if you should make anything else. That's such a tiny bug he looks lost way over there; try putting him on top of something else—or close to something. See—he does look better, doesn't he? Do you think you have too many things in your collage? Yes, they do cover each other too much. Well, that's no problem. Just take out one or several of them; perhaps later you can use the extra ones in another picture.

Have each child clear away his supplies as soon as his collage is finished. Do you have a good title for your collage? Put it on the back of your picture. The children who finished their work before the others should wait—and watch—while the other children complete their work. You can see all kinds of things just by looking around you. Stay in your own seat, of course, so you won't bother the people who are still working.

Give each child an opportunity to show his work to the rest of the class before you arrange a display of all the work. You wouldn't have believed that there are so many things here where we live!

Make It Easy—For Yourself!

1. If "collage" is a new word, explain that it is a French word that means pasting. A collage frequently has several textures or materials in it. If you are working with young children, talk about texture. Explain the meaning and let the children feel the texture of the materials they will be using.
2. No pencils! No preliminary drawing. Children will do better if they are taught to think and then cut. Don't add pencil or crayon details; if they are too small to make from paper, they are too small to include.
3. Have each child take only a couple of the materials at first; this will get everyone started quickly and will make sure each child has a good selection. Encourage them to share supplies with other children who have some left over. They should be allowed to return to the supply area whenever they need larger or different materials than they can get from other children.
4. Have white and several colors of 12″ × 18″ construction paper for the background and a variety of 9″ × 12″ construction paper to cut. Your supply of cloth, sandpaper, corrugated paper, cork—or other textured materials—can be cut smaller to stretch your supply.
5. If you use cloth as one of your materials, it is a good idea to have available some scissors that are to be used for cloth only. Scissors

that are used to cut paper soon become so dull that it is difficult or impossible to cut cloth with them. You might want to keep a second set of scissors that are used with cloth only. Mark the handles in some way (tape, yarn) so they won't be mixed with the regular scissors.

6. As children work and you walk around the room, carry a 12″ × 18″ paper as a tray on which children may place flat scraps of paper they want to get rid of. It will give them more work space and make the final cleanup easier and faster.

7. If paste brushes are not available, have children make paste applicators by folding scraps of paper until they are narrow strips, bending them in the middle to give them added strength.

8. Teach children to do their pasting on newspaper to keep their pictures and work areas clean. Paste around all the edges of an object—not in the center or in dots.

9. Encourage children to give their pictures creative titles; a good title can improve an otherwise ordinary picture. Titles should be put on the back of the pictures where they are handy for display purposes, but don't detract from the appearance of the picture.

10. If you are working with young children, have each one list one new thing in his picture. Put the list on a chart or on the chalkboard and use it as a reading or spelling lesson.

lesson 2
It Can Be Beautiful

OBJECTIVES

1. To make children more aware of their environment and more responsive to it.
2. To encourage children to use found materials as a part of their art.
3. To experiment with organization to create a pleasing arrangement.

You are appalled at the litter around you. Places that should be lovely are made ugly by carelessly discarded junk of all kinds. However—it can be beautiful again!

Have you ever seen someone toss away a candy wrapper as he walked down the street? Have you ever done it yourself?

There will be nods of agreement. Just think what would happen if everyone tossed away their scraps instead of putting them in litter containers! Every place would be a mess, wouldn't it? It really would be just as easy to hold on to that candy wrapper or put it in your pocket until you came to a rubbish can, wouldn't it? Sometimes we get careless and don't think how ugly scraps can make a place.

Have you ever stopped along the roadside for a picnic, then found some thoughtless person had been there ahead of you and left all kinds of litter? It wasn't a nice place to eat, was it? You probably went on and tried to find a cleaner place.

Children will probably have stories they want to tell of someone else's carelessness and their own good deeds. Let them talk for a while—and compliment them for their thoughtfulness: for putting their soda bottles in the rubbish can—or keeping the popcorn boxes until they got home—or picking up the paper someone else had dropped.

Have you noticed the schoolyard lately? There is lots of trash that has been dropped carelessly. Let's do something about it; in fact, let's do two things about it. First we'll pick up some of the scraps to make the yard look better—then we'll make a picture of the things we find.

The children may think that a strange idea—a picture out of junk! That's right! Each of you will make a picture with whatever you find. It will be a litter collage—and you will be surprised how nice it will look!

Explain that this time they are only to pick up the small scraps because they are going to make a collage only as big as the cover of an egg carton; so you won't be able to use anything large. Warn them, too, not to collect anything which is wet or which is unusually dirty; only clean scraps this time.

Give each child a small paper bag and go into the schoolyard. Have a bag of your own so the children will see you are willing to take part in the same activity. You will be able to use your litter for a demonstration later.

Keep the class close to you when you first go outside; roughly define an area within which you want them to stay. Make it large enough so children won't get in each other's way, but not so large that you can't supervise it properly. Tell them they will have only five minutes to make their collections.

See—there's a bottlecap you overlooked; I'll put it in my litter bag. Yes, there's another one for you. Oh, don't flatten the paper; that will make a more interesting shape in your litter collage than if it were flat. You were lucky to find that broken red balloon; it will be an interesting color as well as a different shape. Don't you think that is too big for the collage you are going to make? Just drop it in the rubbish can. Certainly that little piece of rag is fine to save; you will be able to bend it in any direction. Good for you! A piece of string is a wonderful thing to find; probably no one else will have any.

After a few minutes, when you think the children have collected enough litter, call them together and go back to the classroom. Don't take anything out of your bag yet.

Have the children gather around you. Cover a table with newspaper and empty all the things in your litter bag onto it. Yes, some of you found some of the same kinds of things I did; but I don't think any of you found a broken button.

Take the pâpier-maché cover of an egg carton for the background on which to make your collage. The raised edges are like the raised part of the frame on a picture, so when our collages are finished they will be already framed and ready to hang. Did you see the two holes on the edge where it closed over the bottom of the egg carton? All it needs is a string through them.

Begin to arrange litter inside the cover; begin with something which is reasonably large. Overlap it with something else, and add a third thing. Move them about as you add another piece. It begins to look like pretty good litter, doesn't it? That gum wrapper which has come apart will be a good thing to put in that empty space. I'll move that crumpled paper just a bit so it overlaps the wrapper. And look—isn't that a good place for my broken button? See, it attracts attention to that spot because it is different from anything else. Can you find something that would look good in that empty space right there? Good! And now it looks finished, doesn't it? I didn't use all the litter I collected, but I don't have to use all of it—and you won't have to, either. If there is some left after your collage is finished, you can put it back into the litter bag and throw it all into the wastebasket.

Take the arrangement you have made, dump it on the newspaper, and add the leftover litter to it. I'm not going to glue mine to the cover; instead, I'm going to leave here all the things I collected, and if any of you would like to use one or two of them in your collage, you may take them. No, no—not now! Later, when you know whether or not you need them.

Explain that the first thing each of them has to do is empty his litter bag on the newspaper that covers his desk. Choose one thing to add to this sharing supply that I have started; every person will add just one thing. If you only collected three things, you will still have to contribute one of them to this supply. You may select things from it to add to your collage, but you will have to give everyone else a chance to select some- thing, too. Put your litter bag to one side and save it to put the leftover things in before you put them in the wastebasket.

See that each desk is covered with newspaper and give each child the cover of an egg carton and a tube of glue.

You won't glue anything to the egg carton cover until you are sure your arrangement is just the way you want it. That match folder was a good thing to find! Don't you think it would be a better arrangement if some of the things overlapped? If you need something else, get it from the sharing supply. What would it look like if you moved that match so it seemed to be pointing to the peach pit instead of pointing outside the picture? It makes both things look better, doesn't it? Remember I told you that string was a wonderful thing to find? See how you can make it move about in your collage—even under some things and over other things! Glue them in place as soon as you have a good arrangement. No-no-no! Don't take them all off; now you will have to make your arrangement again. Just take off one thing at a time—glue it, and put it back again.

As each child finishes his work, have him fold the remaining part of his litter in the newspaper that covered his desk, put it in his litter bag, and drop the whole thing into the wastebasket. After everyone has finished selecting anything he needs from the group supply, have one child fold it inside the newspaper and discard it.

Glue dries slowly, so it will be better not to have children hold their collages up to show the rest of the class. Instead, let all the children take turns—a group at a time—walking around the room to see all the work. Call their attention to an especially nice arrangement here and there—or to something especially interesting that was found and included in the collage.

Yes, it was just litter—and ugly—when it was carelessly thrown away, but see—it can be beautiful. It is beautiful!

Make It Easy—For Yourself!

1. Know whether or not there is enough litter on the school grounds before you take your class out to look for it. You may have to take a litter walk near the school.
2. Have a small paper bag for each child. The store where you buy your groceries will probably furnish enough of them for you.
3. If "collage" is a new word for your class, talk about its meaning. Explain that it is a French word for pasting (or gluing), and that a collage frequently has a variety of materials of different textures in it.
4. If you have some gold spray paint, you may want to spray some of the collages. Let the children tell you whether they would like to have theirs sprayed or left to show what each part of it is. Spray painting should be done outdoors.

16 The Sun

lesson 1
Shadow Weather

OBJECTIVES

1. To observe the effect of the sun in creating shadows.
2. To learn to observe more closely; to see the things we usually merely look at.
3. To experiment using the sun to create new shapes.
4. To introduce the idea of background space as an important part of a picture.
5. To realize that the varying direction of light makes shadows change shape and size.

You look in the paper every day to learn the weather forecast. Rain, snow, sleet—stormy weather; fog, cloudy—dull weather; *sun*—shadow weather!

Have you ever seen a design on the ground that no one made? No, it couldn't be an arrangement of patio blocks because someone made that. No, it couldn't be a lovely flower garden, either; someone planted that. This has to be a design that no one made. I'll give you a clue—the sun made it.

Someone will almost certainly think of shadows. Bright sun makes shadows of everything! It makes lovely designs!

227

They're Mine?

What kind of things cast shadows? Trees do—and fences, and houses. The sun can make a shadow of you too. What else? The sun can make a shadow of anything—anything that isn't flat against something, that is. Hold up a small thing that is near you. Could the sun make a shadow of this? Sure it can—and there it is.

Would a shadow of this eraser always look exactly the same? Well, let's see. Hold it parallel and close to a flat surface on which it will cast a shadow. Looks the same shape and size as the eraser, doesn't it? Trace around it with a pencil, then tip the eraser. It looks different now! Trace around it, too. And look—it is still different. See how thin it is—and now how fat it is! Each time you create a new shadow, trace around it. Move the eraser each time so the shadows create an interesting pattern over the paper.

Have you ever seen your own shadow? Sometimes it was in front of you, sometimes to one side of you, and sometimes in back of you. Sometimes your shadow was fatter than you and sometimes you were bigger than your shadow! What do you suppose made the difference? Of course! It depended where the light was—or where you were.

What color is a shadow? Children will probably think a shadow is always black. Hold something over a light green paper. The shadow is green—a darker green than the paper. Hold the same object over a light red (pink) paper. The shadow is red—a darker red. Repeat the experiment several times until the children realize that a shadow can be any color—just a darker shade of whatever it touches.

Let's go back to that paper on which I traced the shadows the eraser made. Are there some empty places? All right, let's put a shadow there; I'll have to move the eraser until it makes a shadow that fits the space. See—there is one, so I'll trace that shadow. Fill any other areas of the paper that look empty. Comment about the variety of shapes that all came from just one shape. The sun does very special things, doesn't it?

There's one thing wrong with these shadows, though. Can you tell what it is? Look carefully and I'll show you a clue. Hold the eraser so it recasts a shadow that is already drawn on the paper. How does that differ from when you just see the drawn one? The shadow color was darker, but there is another difference, too. The shadow on the paper is solid, but the one I drew on the paper is just a line. The real shadow looks better, doesn't it? Well, that is easy to fix.

With a felt-tipped marker the same color as the light paper your shadow arrangement is on, go over the shadow. Fill it in so it is solid. Now it is the way a shadow ought to be—solid, and the same color it is on, but darker.

Each child will be anxious to create his own shadow picture. Let each of them choose an object for the shadow and a piece of light-colored construction paper; then let him find a sunny place and go to work.

A good way to begin is to make the shadow look just like the real thing. That's the way to do it—just trace around the shadow. Hold it still or you won't be drawing the shadow. Can you make it look longer and thinner than it really is, or shorter and fatter? Don't let any of the shadows touch or overlap, because you are going to make them solid and all the same color later. What color will your shadows be? They will be the same color as your paper, but they will be darker. Yes, your shadows will be red—but yours will be blue. Have you tried tipping your object so it creates a long, thin line? Your shadows are all in the same direction. Try moving your paper so you can make some shadows that go different ways.

When the children have finished drawing the shadows, give each one the color felt-tipped marker he will need to complete his picture. Continue to walk around the room to help in any way you can. Compliment the child who is working carefully. Show another child how to get smoother edges on his shadows by drawing a marker line around them first. See—isn't that better? Shadows have as smooth an edge as the thing that made the shadow.

Every child will be guaranteed success. It will be a proud and exciting time as they are shown to the rest of the class. Later, when they are on display—and the room is filled with shadows—you won't have to look for the weather report. You'll know it says sunny—but you'll know it's shadow weather!

Make It Easy—For Yourself!

1. Pick a day when the sun is bright and casts clear, crisp shadows.
2. You may want the children to go outdoors to draw the shadows. Tape the colored construction paper to a heavy piece of cardboard to have a good drawing surface—and so a slight breeze won't move the paper.
3. Try the markers on lighter paper of the same color before you have the children use them, to be sure there will be enough contrast. Avoid yellow markers—they are too light.
4. Use either 9″ × 12″ or 12″ × 18″ light-colored construction paper. Older children will want to use the larger paper; the smaller size will be better for younger children with a shorter interest span.
5. Be sure not to use any more of one color paper than the number of markers you have of that color.
6. Have a variety of items from which children may choose: spoons,

twigs, large buttons, scissors, candles, small books, toy trucks—or let the children find or make their own shadow items. Tape a tongue depressor or applicator stick to the item if it needs a handle to hold it in order to trace the shadows better.

7. When the class is outdoors, keep them reasonably close together so you will be able to supervise their work. Define an area to be used and let them choose their own spot within it.

8. Return to the classroom to use the felt-tipped markers. Cover the desks with newspaper since the ink from the markers may soak through the construction paper.

9. The wide markers are more satisfactory to use for this lesson than the fine line markers.

10. If felt-tipped markers are not available, use black crayon, ink, charcoal—but be sure the children understand it is to create a strong contrast of color and that shadows are not necessarily black.

I like to lie in the sun under a beach umbrella.

lesson 2
Fun in the Sun!

OBJECTIVES

1. To use pleasant and personal experiences as motivation for an art lesson.
2. To provide opportunity for working with a fluid material that makes a rapid expression possible.
3. To translate an idea into visual form.
4. To establish methods and techniques which can be used in future art lessons.
5. To learn to paint large and to fill a paper without crowding it.

It has rained for days and days—but now the weather has cleared and the sun is out! You can't wait to get outside and enjoy it. You want to have fun in the sun!

Have you ever had to stay in the house for a long time? Perhaps you were sick—or maybe it had rained or snowed for several days, and your mother wouldn't let you outside. Then, when you were feeling better—or the sun came out—your mother said it would be all right to go outdoors. What is one thing you would like to do?

Children will have an immediate answer, but it may be general and indefinite. They like to play. Ask questions that encourage children to expand their thinking. Do you like to play by yourself or with a friend? What do you like to do together? If you wanted to play football, it would be better if you played with several friends, wouldn't it? Yes, it is fun to ride a bicycle. What color is your bicycle? Where do you play hopscotch? How would you show us that in a picture? What do you like to do outside on a cold, winter day? How big would the snowman be? Taller than you? You'd have to make him very big on paper, then; you'd have to reach up as high as you could to put the hat on him.

Encourage the children to think of many kinds of things. Talk about things they like to do in the sun at different seasons of the year. Make comments and ask questions that will help them see details. Where were you? You would only do that during summer, wouldn't you? What could you do in a picture to show us it was summer? Were you by yourself? How could you show a crowd of people? How did it feel? What do your arms do when you are skating? The more ideas the children have before they begin to paint, the more valuable the experience will be and the better and more personal their pictures will be.

Organize the class into groups of three to five children. Have an extra desk for each group where supplies may be shared by the children. A palette of paints, a can of water for washing brushes, and a paint brush for each child should be at the sharing area.

You have a big paper, so put something big in it first. You are the one who is having fun in the sun, so make you big. You can make yourself as tall as the paper and still have space for other people and other things. You aren't the shape of the paper, so of course there will be more space, and you can be standing partly in front of some things. If things are farther away, they look smaller than you, so there are lots of reasons why you can be big in your picture. Good! That's a fine beginning! We can see that you are having fun roller skating. Can you paint something now that will tell us where you are? You do need a sunny day for a sailboat ride, don't you? I can even tell that you are in your bathing suit. Oops— don't let any paint drip onto your picture. Wipe the brush off once after you dip it into the paint. Can you think of any way of using that paint spot so it won't show in your picture? Yes, that would be a good thing to do. Good for you! No one else thought of having a picnic! Did you cook over a campfire? Don't put too much in your picture. When it is nicely filled, stop and don't add another thing. An artist has to know when to stop, you know.

After several of the children have finished their paintings, urge the other children to finish quickly. Think of just one more thing that will complete your picture. Some children become so involved in the painting process that they would go on and on painting without much thought involved. It is better to stop sooner and to paint more frequently.

Children will want to share their pictures and ideas with the rest of the class, but clear away the supplies first. This will give the paintings a chance to dry so it will be safe to hold them up to be seen. Encourage the children to tell something about their action, how they felt, how it came about. Tell something that makes the paintings come to life rather than merely identifying the objects.

Your room will be a happier place than it was before. You can't help but be happy when you are having fun in the sun!

Make It Easy—For Yourself!

1. Take plenty of time to let children express their ideas verbally. Draw out details of color, size, action, and location that will make their ideas visual.
2. Cover all work areas with newspaper, including the sharing desk.
3. If you do not have enough desks to provide for sharing desks, let some children work at easels, tables, counter space—even on the floor. In that way there will be enough extra desks to use as sharing areas.
4. Small cans, about the size of soup cans, are a suitable size for washing brushes. Fill them about half full so water won't splash or spill as brushes are washed.
5. Egg cartons make fine palettes. They can be broken in half to provide space for six different colors for children to share. In this way, even the youngest children can carry all the supply for a group at one time. They are stable and do not tip over easily. If you want the children to have individual palettes, break the carton into thirds.
6. Plastic squeeze bottles make fine paint dispensers if your paints do not come packaged that way. Paint left in them remains moist and ready to use from one lesson to another.
7. Assign one child from each group as a helper. He will take care of the supplies for the children in his group: (1) cover all desks with newspaper; (2) give each child a piece of 18" × 24" newsprint (or white drawing paper for older children); (3) put an easel brush for each child on the sharing desk (or large wash brushes for older children); (4) put half a can of water on the sharing desk; (5) put an egg carton palette of paint on the sharing desk. (If individual palettes are to be used, let one group

of children at a time come to the supply area for their own paints.)

8. Be sure children stand to paint. It provides greater freedom of motion than when they sit, and so results in better work as well as fewer accidents. Teach children to hold their brushes near the top end, since this permits greater movement and, therefore, results in larger and freer pictures.

9. No pencils! No preliminary drawing—just paint.

10. Children should sit as soon as they finish painting. This lets you keep track of who is still painting, and lets those who haven't finished have a chance to complete their work without interruption.

11. The final cleanup is easy if well organized. The same helpers should: (1) collect the brushes and leave them on a newspaper at the sink so they can be washed later; (2) bring the cans of water to you at the sink to be rinsed and left on a newspaper to dry; (3) bring the sharing palette to you (empty, stack, wrap in newspaper, and put them in the wastebasket); (4) fold the newspaper on the sharing desk and put it in the wastebasket. Last of all, each child should stand, slide the newspaper from under his painting, fold the newspaper two or three times, and pile it on the sharing desk so the helpers can put them in the wastebasket. If individual palettes were used, have them put on the sharing desk. Go from group to group, stacking and wrapping them in newspaper so the helper can put them in the wastebasket.

12. Brushes should be stored flat or left standing on their wooden ends.

17 The Moon

lesson 1
Moon Play

OBJECTIVES

1. To use interest in moon exploration as motivation for an art lesson.
2. To experiment with a single shape to create an over-all pattern.
3. To develop ability to arrange parts to create rhythm and balance.
4. To devise ways of creating variety within unity.
5. To learn to use a compass.

Sometimes you feel as though you're going around in circles! Or is it just moon play?

Have you ever looked up on a night when there was a full moon? It looked like a huge orange circle in the sky, didn't it? Remember seeing pictures of the surface of the moon? It was covered with craters—circles over the surface of the moon. Some of them were so big—so deep and so high—they were more like round mountains or valleys, while some of them were just little circles on the surface of the moon. Circles are important to the moon, aren't they?

Explain that they are going to make pictures that will just be made of circles. They won't really be pictures of the moon, but we'll use that for our idea.

Moon, Moon, Moon

It wouldn't be very interesting, though, if every circle were exactly alike, would it? They wouldn't all be alike even on the moon. Circles are all the same shape, but what are some ways in which you could make them different? They could be different sizes, just as the craters on the moon are different sizes. How else could they be different? They can be different colors. Show the class the variety of colored construction paper that is available. That will make them more interesting, won't it?

Cut a fairly large circle from a piece of paper. I can't change the color of this—and, of course, I want it this size, or I wouldn't have made it this big. Aren't there still some things I could do to this circle to make it differ from another one this size and color? Cut a piece out of the side of it? No—it might be the shape the moon sometimes appears to be, but it wouldn't be a circle. This time our moon pictures are going to be nothing but circles. You could overlap two circles so it looked like a piece was out of one of them. What else? A doughnut is a circle, but it has a hole in it! Well, let's make a hole in this. Cut a medium-sized circle from the center of the large circle. Would it have looked different if I had cut just a tiny dotlike circle—or a much bigger one? Suppose I hadn't cut it from the middle but had cut it off-center; that would have made it still different. I could have cut it so it was just a narrow band.

All this time we have been talking about cutting circles. Is there any other way of making a circle out of paper beside cutting it? Of course—you could tear the circle. That would make it different from the others.

Will you want to do all these things in one picture to make circles look different? No—that would be too many things different. Perhaps a good way to start would be to decide which colors you would like to use. You may decide to use just one color and have the circles different in several other ways, or to use two, three, four, or even five colors. Do you think it would look right to have them all the same size? Well—they could be, or you may want to make them several sizes.

Have you ever used a compass? A compass is a tool that helps you make a circle. Let's see how to use one.

Let the class gather around you while you demonstrate. Have you ever noticed that new compasses come with short pencils in them? Why do you suppose that is? Even if you put a pencil of your own in the compass, it would be a good idea to use a short one; we'll see why in a minute. Adjust the pencil so the tip is slightly below the sharp point of the compass. Spread the pencil and point apart of couple of inches. Stick the point of the compass into a piece of construction paper (it is a good idea to have a cardboard or several thicknesses of paper under it so the point won't make a hole in the furniture). Take hold of the small, round projection at the top of the compass with your thumb and forefinger, and roll it between them so the pencil traces a circle on the paper.

What would happen if there was a long pencil in this compass? Make another circle. If it were a long pencil it wouldn't go under my hand as I turn the compass—so I wouldn't be able to use it correctly.

How would you make a smaller circle right in the middle of this large one? Just push the pencil closer to the point, stick the point back into the same hole, turn the compass—and there is a smaller circle that could be cut out to make a doughnutlike shape. What would you do to make a smaller circle off-center in this big one? Someone will see that they should move the point of the compass to one side.

Have two or three colors of 12″ × 18″ construction paper for the background, including one dark and one light color, and a wider variety of 9″ × 12″ construction paper for the circles. Let each child select a large paper and one or two smaller ones. Later they will almost certainly need more colors—or additional pieces of the same color. They may be able to share leftover pieces with other children, or they can get what they need from new supplies, but this way everyone will be able to get started in a hurry. While they are selecting materials, give each child a compass and a pair of scissors.

That's a good way to begin—by making a large circle first. Now you can use the rest of that paper to make different size smaller ones. Lay them on the large paper right away to see what they look like. What will happen if you make circles the same color as the background? They won't show because there won't be any color contrast. You could do that—put them on top of another color. Then they would look like a smaller circle had been cut out so the background paper showed. You just have separate spots on your picture so far; wouldn't it be a good idea to over-lap some of them—or to place some of them closer together, so they make groups of circles? Try it. Are all the large circles on the same side of your picture? Doesn't it make that side too heavy? Well, that's easy to change—just move them until they balance better before you paste any of them to the paper. That's an interesting arrangement! It looks as though the circle is spinning on and on. Have you made several kinds of things different about your circles? Looks good, doesn't it! Paste them as soon as you have an arrangement that balances and looks well on the paper. Paste all around the edge of each circle so it will lay flat and smooth on the paper. Do you have an empty space there? Perhaps one more small circle would finish it.

This is an activity in which every child can be successful. Encourage, assist, and compliment each child until he has done his best and is proud of his work. He will be anxious to show it to the rest of the class—and later, to see it on display. Encourage the other children to comment about unusual variations, pleasing color contrasts, rhythmic and balanced arrangements, and good workmanship.

No, you're not on the moon—and you're not really going around in circles, either—but it is interesting and pleasing moon play.

Make It Easy—For Yourself!

1. Teach children the correct handling of a compass. Whenever a new tool or new material is introduced, children should be taught its characteristics and the right way to make use of it.
2. Do not use compasses with younger children; let them cut the circles freehand. Don't worry that they aren't exact circles.
3. Give each child a half page of newspaper to do his pasting on; it can be folded in half later for another clean area. Teach children good work habits—they will do better and keep their work area clean, too.
4. Don't be in a hurry about giving children paste. If they have it at the beginning of the lesson, they are apt to paste each circle as soon as it is cut. An important part of the learning experience is the arranging and rearranging of the design to create a rhythmic and balanced picture.
5. Satisfactory paste applicators can be made if paste brushes are not available. Fold a scrap of paper several times until it is a narrow strip, bending it in the middle to give it added strength.
6. As you walk around the room, carry a 12″ × 18″ paper as a tray on which children may lay flat scraps of paper they want to discard. It will give them more work space and make the final cleanup faster and more orderly.
7. When you see children have finished with their scissors, collect them as you walk around the room.
8. Encourage children to think of interesting and creative titles for their pictures. Have them put on the back so they will be available for display purposes but won't detract from the picture.
9. Arrange an exhibit of all the pictures. Each child will feel pride in his work and want to see it displayed.
10. Encourage the children to bring in pictures of the moon and its surface. Arrange a display of them—perhaps mixed with their moon play pictures. Encourage the older children to use the library to find additional information about the moon and its exploration.

lesson 2
Trip to the Moon

OBJECTIVES

1. To capitalize on interest in moon exploration as motivation for an art lesson.
2. To provide an opportunity for children to choose either an individual or a cooperative project.
3. To experience painting on a three-dimensional surface.
4. To use polymer as a paint medium.

Four—three—two—one—blastoff! You're on a trip to the moon!

It's exciting to watch a rocket blast off on its way to the moon, isn't it? How would you like to be one of the astronauts inside? Well, maybe someday you will be—or you might be a passenger going along for a vacation on the moon.

Talk about some of the things they would see on the ground before blastoff, on the way to the moon, after they landed on the moon, on the way back to the earth, during splashdown, and during recovery. Encourage each child to add something to the conversation.

The rocket stands on end and is attached to a tower before it blasts off. It is extremely tall, isn't it? Which part do the people ride in? That's the only part that will return to earth at the end of the trip, so a picture of the rocket on its way to the moon would look very different from the

Coming Home

rocket on its way back from the moon. Does the rocket land on the moon? No—only a part of it. Does all of that part return to the rest of the rocket that has been circling the moon? No—and even that part would be discarded in space after you were safely back inside.

What do you suppose the moon would look like as you approached it? What would the surface of the moon look like when you stepped out of your landing craft? What would you look like? Would you have any kind of vehicle with you that would help you move about on the moon?

Continue to ask questions about various stages of a trip to the moon and back again. Encourage children to describe details, color, and unusual sights. Ask questions and make comments that will help them express their ideas as visually as possible.

When you come home from a trip you like to bring pictures of it with you. It's fun to look at them yourself, and it's nice to be able to show them to other people. Imagine how important pictures of a trip to the moon would be! You'd want something to show every part of it, but a trip to the moon is much too important to just have the ordinary kinds of pictures. Your trip-to-the-moon pictures are going to be made so you can see them all at the same time.

Show the class one of the larger boxes you have. They'll be made on boxes like this. How many sides are there to a box? Are you sure? Count them. There are six sides, so you can have how many pictures? You can't have more than five pictures. You won't leave one side empty. There wouldn't be any reason for doing that, would there? That must mean, then, that one picture is going to be on two sides of the box. You won't paint the same picture on two sides; you will make one picture so big that it will fill all of one side of the box and go right on to another side and fill that, too. It will be an important part of your picture collection; it will be like an enlargement.

Could you make more than one picture an enlargement—make it take up two sides of a box? Yes, if you want to. On some part of your box a picture will extend from one side of the box right around and onto another side. On another part of the box one picture will fit just one side. Not everything will extend from one side of the box to another, but two of them may. Could you have three pictures extend from one side to another? Are you sure? Let's see. Three of them would need to have six sides, and there are only six sides to the box, so that wouldn't leave any side for one whole picture. So no, you can't make three enlargements.

Show the class the assortment of boxes you have—some small ones only a few inches in each direction, others the size of a shoe box, others giant size boxes. Have a variety of shapes as well as sizes—some almost square, others thin but wide across, others much longer than they are wide. Explain to the class that they may work alone if they like, or two children may work together on one of the larger boxes. Remind them that if they decide to work with someone else, they should be sure to choose a person they will work well with—someone who works much the same as they do and someone whose work they respect.

Let any child who wants to do so choose a partner and then select one of the larger boxes. Partners will need a few minutes to talk over ideas and make their plans. Also see that they have work space enough for the large box they have chosen. Assign a large table to them—or a counter space, or move two desks together. While this is being done, let those children who are going to work alone choose their boxes. Their own desks should be enough work space for them.

Stop the class and explain about the painting. They will use regular tempera paint, but they will add something to it to give the paint a hard, shiny finish when it is dry. It will look almost like enamel paint—or as though you had varnished the paint. To make it look that way you will add just a bit of polymer to the tempera paint as you use it.

If your class has used polymer before, they will remember that it is a synthetic material; they probably used it as an adhesive. If this is the first time they will be using polymer, show them the white liquid and explain its uses—as an adhesive, and as a medium to mix with paint to change the quality of it. Remind them that because it is a powerful adhesive, there are certain precautions they must take. The paint brushes must be wet before they are put into the polymer, and they must be kept wet all the while they are being used—and until they are thoroughly washed. Have a can of water in which to drop the brush; it must stay there any time you are not actually painting with it. If even a drop of polymer is spilled on clothing or furniture, it must be wiped off immediately with a damp sponge. It does no harm as long as it is wet, but it is waterproof once it is dry.

Each child—or each pair of partners—will have half an egg carton as a palette. One section will be for a supply of polymer and the other five may be used for paint. You may have five different colors, or you may want to save one or two sections for mixing new colors; you may decide that when you are ready for your supplies. Mix a bit of polymer into the color in the palette, or just dip the wet brush into the polymer each time before you put it into the paint.

Place several cans of water in different parts of the room so every child will have easy access to one of them. Give each child a brush and see that it goes into the can of water immediately. When plans are made, let children select their colors. It will get things started quicker if you pour the paint into the palettes the first time. Older children will be able to get additional supplies as they need them; you may want to continue to dispense them to younger children.

Each child will soon be hard at work. Your help will be needed everywhere at the same time, so move quickly from child to child and group to group. Ask questions to help children think through their problems; answer questions when necessary. Encourage those children who are slow or hesitant. Compliment partners who are working well together, and help others settle disagreements as they arise.

You both have to agree on what you are going to paint. When you work with someone else you have to talk things over so both people are satisfied. Keep that brush in the water all the time you are not using it. No, the polymer won't lighten the color; it won't change the color at all— it will just make the paint have a shiny finish when it is dry. That was a good part to make big enough to take two sides of the box. You must have seen the capsule coming down with the parachutes open on television; you have made a fine picture of it. Oops—wipe up that little bit of polymer before it begins to dry; just a damp sponge will do it. You'll have to wait for one side to dry completely before you can turn it over to paint the last side of the box. You have a wonderful selection of pictures to bring back from your trip.

When the boxes are finished, they can be displayed in a variety of ways. Small boxes can be hung as mobiles. Medium-sized ones can have a rod inserted from one corner through the box to a diagonal corner; then the rod can be inserted in a can of sand or nailed to a wooden base. Giant-sized boxes will probably be left as they are and displayed on a counter— or standing on the floor if they are extra tall. The variety of sizes makes a variety of display techniques appropriate.

They will make conversation pieces for a long time to come. Well, why not! It isn't every day you blast off on a trip to the moon!

Make It Easy—For Yourself!

1. No pencils! Children will do much better work if they think and then paint without any preliminary drawing.
2. You will probably have to take two lessons to complete the work. It takes time to plan—and remember, there are six sides to paint.
3. If polymer is not available, use tempera paint without it. If you like, you can have the finished paintings varnished to give the paint a smooth, glossy finish. Remember to clean varnish brushes in turpentine.

4. Have some extra large brushes available for painting extra large surfaces.

5. Cover all work areas with newspaper to protect them from the polymer. If any of it spills, wipe it up immediately with a damp sponge. It is no problem while it is wet, but it dries quickly and then is waterproof.

6. Children should wear smocks to protect their clothing. A man's large shirt (with the sleeves cut off near the shoulders) worn backwards makes a fine smock.

7. Don't spread the paint too thin. If it is applied rather heavily, it will cover any advertising or designs on the boxes. If polymer is not added to the paint, you may have to add a small amount of liquid soap to the paint to make it adhere to shiny surfaces.

8. Dip brushes into water before putting them into the polymer. Keep them in the water until they are *thoroughly* washed in cold water. Wash them several times to remove every trace of polymer. Polymer is an adhesive, and if any of it remains in the brushes they will be ruined.

9. Polymer on your hands reacts much like glue. It will not wash off; the polymer will peel off in filmlike pieces. It helps to rinse your hands in warm water—but soap does not help.

18 The Stars

lesson 1
Twinkle, Twinkle Little Star

OBJECTIVES

1. To demonstrate the importance of repetition to create rhythm and balance.
2. To use a familiar material—chalk—in a new way.
3. To introduce positive and negative forms of a stencil.
4. To introduce a simple stencil technique.

"Twinkle, twinkle little star . . . Like a diamond in the sky."

You've looked up into the sky on a clear night, haven't you, when the sky is filled with stars. Every one of them twinkles and shines, but some of them seem to be brighter than others; some seem to be bigger than others. Do they have five or six points on them the way we are used to seeing pictures of stars? You can't see points at all, but they seem to be alive with light that dances and moves. I guess that's why we usually put points on them.

Children will want to comment, so let them talk about stars for a few moments. Ask a question or add something which will help visualize the comment.

Let's be different with our stars today. They may have points on them, if you like, to make them twinkle, but let's not make the ordinary

Twinkle at Twilight

five- or six-pointed kind. Fold a piece of white paper into quarters. Leave about two inches along the edge, and from the long folded edge cut a long, narrow point. From the shorter folded edge (leave about two inches) cut another narrow point so the two cut lines meet near the folded corner. Open the paper. See, a shining four-pointed star. Fold another paper. Are there other ways I could cut points into this paper that would be different from the first one? I could make more points—or make them different shapes—or sizes. Would you like to try making a different kind? Be sure you cut from the folded edges toward the center fold. Good! That should be different. Open it carefully and let us see what it looks like. See—isn't that a shining star—but not like the kind you usually see! Who would like to fold another paper and make another different star? The paper can be folded differently, too.

We're going to use these stars as stencils. What is a stencil? A stencil is a shape that can be reproduced—another one made just like it—over and over again by rubbing color through it or around it. We could rub color around the stars we have just cut, but we couldn't rub color through them, so there must be another kind of stencil.

If we are going to rub color through it, there must be a hole in the stencil. The piece of paper that was left over after we cut each star stencil has a hole in it—a hole that is the same shape as the star we cut out of it. That is a stencil, too—it is a stencil that we could rub color through to reproduce the same shape.

These shapes are different, and so have different names—just as people have different names. The solid shape is called a positive stencil; the one with the hole in it—the one with the shape *not* there—is called a negative stencil.

Now let's see what we can do with these positive and negative star stencils. Lay a negative stencil on a piece of newspaper and draw a rather heavy chalk line on top of the stencil, close to the edge of the star. Lay the stencil on a piece of 12″ × 18″ white drawing paper. Wrap a piece of facial tissue around your index finger, and use it as a brush to rub the chalk from the stencil onto the white paper. Hold the stencil firmly in place as you make the rubbing motions close together so they make a solid color around the edge—not stripes. Lift off the stencil carefully to avoid smudging the chalk. There it is—a chalk reproduction of the negative stencil. Pretty, isn't it!

Comment about the shape that was created. It is the same shape as the stencil, but there is one thing that is different. The negative stencil was solid around the star, but the chalk has made a solid star with nothing around it. Just the opposite, isn't it! What do you think will happen if we use the positive stencil to make a star? The star will be just white paper with color around it—a negative shape. It will be just the opposite of the stencil.

Draw a chalk line near the edge of the positive stencil. Again, lay it on the white paper, wrap your index finger with the facial tissue, and rub outward so the chalk is pushed from the stencil onto the white paper. Push the color well out onto the paper into a softly blended edge. Remove the stencil carefully—there will again be expressions of surprise and delight. They look better than you thought they were going to, don't they?

You wouldn't want just two stencils on a picture; it would look empty. Well, you could use the same stencils—or others, too, if you liked —over and over again. Could you overlap them? Certainly—why not? Overlap a rubbing made by the negative stencil with the positive stencil. What would happen if I rubbed it this way? They would be overlapped, but would you be able to tell which one was on top? Are you sure? Let's try it.

Rub all around the stencil—then remove it. You can see all of both of them as though they are transparent. What would you do to make the positive stencil look as though it is in back of the other one? Rub the stencil only to where it overlaps the other one—then stop. You can make many variations to the stencils by the way you use them.

Give each child a piece of paper and a pair of scissors. Think of some way of cutting an interesting but simple star—one that is different. Remember to cut from the folded edge toward the center fold. Oops— are you cutting from the folded edge? What would happen if you cut from the open edge? Your star would fall apart. Yes, it would be a falling star, all right! Cut smooth edges because you are cutting two stencils at one time. Good! That is a different way of making a simple star.

Would you like to make a second set of stencils? Then you can use whichever ones you want to—or both sets, if you like. Don't cut too close to the negative edge—that's where you put your supply of chalk, and you don't want it to tear or smudge the chalk on the outside when you rub it. If you make your second stencils a different size than the first ones, they will make a more pleasing combination to use together.

If one or two children have cut their first stencils incorrectly before you saw their mistake, give them an extra paper and watch them carefully to make sure they are successful on the second try. Once a child has cut one correctly he will be able to make more of them.

Give each child the materials he will need to make the chalk rubbings: newspaper, chalk, facial tissue, and 12″ × 18″ paper. Have both black and white construction paper and let each child decide which he would like to use. Remind them that bright or dark colors should be used on white paper and light colors on the black paper.

Start with whichever stencil you like. Wouldn't it be better to plan the arrangement as you go along—and not jump around the paper? Then things will fit together. Good! That's nice smooth rubbing. Oops—hold the stencil still so it won't move. Your two different size stencils make an interesting variety, don't they? Oh, don't put things in the middle and each corner! You can make a more pleasing arrangement than that. It would look fine if some of the stars went off the paper; just make them do it on several or all the edges of the paper. It will look like a sky full of stars, won't it? Are you finished? Fine! Do you have a good title to put on the back of your picture?

Take time to let each child show his work and tell the title of it. Talk about the good qualities of each. Make each child feel successful and take pride in his work; then display all of them.

You'll think you're looking up into a starlit sky. You'll be reminded of the familiar nursery rhyme—Twinkle, twinkle little star . . . Like a diamond in the sky.

Make It Easy—For Yourself!

1. No pencils! Visualize the shape and then cut it.
2. It is not important—especially with young children—to stress the names "positive" and "negative." It is important, however, to show that there are two stencil forms that can be used differently.
3. Leave an inch or more around the edge of the negative stencil so it won't tear or let the chalk smudge on the outside of the stencil.
4. Cover all work areas with newspaper before using the chalk. It is easier to keep the desks clean than to have to wash them later.
5. Children should wear smocks to protect their clothing. A man's large shirt—with the sleeves cut off near the shoulders—can be worn backwards for a fine smock. Make sure all dress or shirt sleeves are pushed above the elbows.
6. Use 9″ × 12″ paper to cut the stencils. This will allow for large ones and, of course, smaller ones can be cut, too.
7. Always rub the chalk only away from the stencil to prevent the chalk from going underneath it.

8. Paper can absorb only a limited amount of chalk; the rest will accumulate as dust on top of it. Blow it away gently or drop it onto the newspaper.
9. A small piece of scrap paper to rest fingers on while holding the stencils in place on the picture will keep dirty fingers from smudging the picture.
10. If pictures are to be displayed where fingers or clothing could rub against them, you may want to spray them with a light coating of fixative. Hair spray makes a fine substitute for fixative.
11. Simplify the lesson for young children. Let them trace around the stencils with crayon and fill in the shapes, or just use crayons to draw various kinds of stars.

On the first Fourth of July the people attached flat stars to firecrackers and shot them into the air. They went so high that they never came down again. Today we see them all over the sky.

lesson 2
How I Wonder What You Are!

OBJECTIVES

1. To combine creative art with creative writing.
2. To use cut paper to illustrate an original story.
3. To improve ability to plan an interesting and balanced picture.

'Twinkle, twinkle little star, How I wonder what you are!"

Do you know what a legend is? A legend is something like a fable or a fairy tale—it is a made-up story. Legends are short stories that have been told to generation after generation, and they frequently explain how a thing came about. Usually there is some part of them that seems as though it could really have happened, and other parts that don't seem reasonable at all.

For example, there is the legend about a real mountain named The Sleeping Giant. According to the legend, there was an old Indian chief who was very fond of oysters. One day he was especially hungry for oysters, so he walked to the shore. All day he ate oysters. Late in the afternoon he walked back toward home, but he had eaten so many oysters that he became very tired. He lay down and went to sleep. He never woke up and finally he turned to stone. The Sleeping Giant Mountain is that old Indian chief who went to sleep. Part of the mountain is shaped like his head and chin, then there is his chest and stomach, and finally his legs.

Point out that part of the legend of The Sleeping Giant is realistic. There could have been an old Indian chief who was so fond of oysters that he ate too many of them. They could have made him sleepy and he could have lain down and gone to sleep. What are the parts that are not reasonable? No, he probably wouldn't have been able to eat oysters all day. No, certainly he wouldn't have been as big as a mountain. No, he wouldn't have slept that long and turned to stone. But it is an interesting short legend that explains why a mountain looks like a sleeping giant, isn't it?

If you were going to illustrate that legend, what would be one picture you might make? There will be many suggestions: the Indian chief with piles of oysters near him, the shells discarded on the beach; the chief trudging home, so fat he could hardly walk; the chief lying down, dreaming of still more oysters.

Legends are frequently about natural things such as mountains, rivers, the moon, or clouds. There are some legends about stars, too; perhaps you have read or heard about some of them. You are going to make up new legends about stars.

There will be questions, so take time to answer them. It may be a legend about one particular star, or a group of them, or all of them. Just don't let it be anything like any other legend you know. You will have to write a legend first, but don't worry about the spelling or the details now. Just put down on paper the important points; later you will be able to go over your legend to add to it or to correct it—and then to make a neat copy of it. For now, just put down the important things you need to know in order to make a picture of it. How will you get started? Easy! Just think of something very common—like that Indian chief who especially liked oysters. Nothing unusual about that, is there? Then do something about that ordinary situation—he went to the shore to eat some oysters. A perfectly sensible thing to do, wasn't it? Then begin to have something strange, something unreal happen. The old Indian chief ate oysters all day—so many he couldn't stagger all the way home. Then have another strange thing happen that ends with your stars. The chief slept forever and became The Sleeping Giant Mountain! See how easy!

Explain that each child will make one illustration for his legend, and he will make it out of cut paper. Have black, blue, and white 12″ × 18″ construction paper available for the background—and an assortment of

9″ × 12″ white and several colors for the picture.

Give each child a small piece of paper to write his legend on. Take a minute to think—first of something reasonable, then let your mind wander to something imaginary—and then the result. If you make a mistake, don't worry about it. All you are doing now is getting your ideas on paper so you can make a picture.

Walk about the room to help the children. Some children may be hesitant and insecure. Just think of something that could happen: perhaps a mother goes to the store to buy some soup, or a boy is helping his father rake leaves, or a little boy bought a balloon from a clown—or anything you think of.

Some children will want to know how to spell a word—or continually erase to make changes. Don't worry about spelling as long as you can understand it; just draw a line through it if you don't want it there. Work quickly just to get the main ideas on paper.

Other children will write a legend without hesitation. That's a fine legend! Have you decided how you are going to illustrate it? Good! Then get whatever you need and start it right away.

As soon as you see that a child has the essentials of his legend, let him choose his paper and go to work. Scissors will be the only other material he needs immediately.

Good! You chose excellent colors for contrast. The yellow and red will show up fine on the black background. Will there be one thing more important than anything else? Don't you think you should make it first? It is a legend about stars, so I'm sure you could have some extra ones in the background. Your illustration goes nicely with your legend. One of the stars still looks a little bit like a person, doesn't it? Have you tried moving the parts of your picture to see if you can arrange it better? Does it look too crowded there—and too empty in another place? Are you ready to paste your picture?

Continue to help children while they arrange and then paste their pictures. Paste around all the edges of each thing so it will lay flat against the paper; you don't want your picture to look as though it is coming apart. Do you think one more star would keep that place from looking empty? Good! That is just right!

When the pictures are finished, let a few children tell their legends as they show their pictures. When the legends have been perfected and rewritten, have other children read theirs and show their pictures—until all the legends have been heard and all the pictures seen. You will be surprised and delighted with the originality of each child's work. Let them know it, too.

Twinkle, twinkle little star, now I *know* what you are!

Make It Easy—For Yourself!

1. See that each child has the essentials of a legend before he begins his picture. Don't comment about spelling, punctuation, or any of the other language details; the thought is the only important thing in the beginning.

2. No pencils for drawing! Think and cut the parts without preliminary drawing.

3. Don't be in a hurry about giving paste to the children. If they have it at the beginning of the lesson, they will paste things as soon as they cut them. You want them to move the parts around for the best arrangement before anything is pasted.

4. Keep fingers out of the paste. If paste brushes are not available, teach children to make paste applicators by folding scraps of paper several times until they are narrow strips, bending them in the middle to give them added strength.

5. As the children work, carry a piece of 12″ × 18″ paper as a tray so they may lay on it flat scraps of paper they want to get rid of. It will give them additional work space and make the final cleanup easier.

6. If you are working with very young children, let them make their pictures first and then tell their stories. Write them down as they tell them; later, you may be able to put some of them on charts to use during a reading lesson.

7. Plan a language lesson the same day or the next day when the children will correct, improve, and recopy their legends.

8. Encourage the older children to use the library to find legends about stars—or other natural things. Have some of them read to the class or leave them where other children may read them.

19
Other Planets
and Space

lesson 1
Bugs Can Be Good

OBJECTIVES

1. To use interest in space exploration as motivation for an art lesson.
2. To encourage children to want their art work to differ from that of other children.
3. To improve children's ability to create with cut paper.

Bugs are a nuisance! You spray them with bug killer—you'd like to get rid of all of them, wouldn't you! But wait a minute! Did you ever think that bugs can be good?

When you go on a long trip, do you like to bring something back with you to remind you of it? What kinds of things have you brought home from your trips?

Children will want to tell about things they have brought or had given to them. They will vary greatly, and each child will probably try to outdo the one before him.

Those were all interesting things you brought back—and some of them came from very long trips, didn't they? But today you are going to take a longer trip than anyone else has ever taken. No, you won't be going to the moon—not many people have ever been to the moon, but a few

269

Keep people from stealing.

astronauts have been there. You are going farther away than the moon—to a place no one has ever been.

Someone will probably suggest that it may be Mars. Right! You might be going to Mars, or you might even be taking a trip to one of the other planets. You might be taking a trip to Mercury, Venus, Jupiter, Saturn, Uranus, Neptune, or Pluto. That would be an exciting and dangerous thing to do; right now, the only way we can get there is with our imaginations.

Because no one has ever been to any of these places, we can't be sure just what we might find there. There might be people there or there might not be any people. There might be animals or birds or plants —or maybe none of them, either. But we want to bring something special back with us; when the astronauts came back from the moon they brought some rocks with them. Our trip is longer and even more important than the trip to the moon, so we have to have something more important than rocks. We are going to bring back some bugs!

No, they won't be ordinary, pesky bugs. We have too many of them already. These will be bugs you like to have around because they do good things. We'll pretend there are all kinds of good bugs on those planets, bugs that could do good things to help us.

You can only bring one bug back to earth with you, so choose carefully which one you want. If you could have your own private good bug, what kind would you like? What would you like to have him do for you— or for other people?

At first the responses may be slow in coming and those that are suggested will be unimaginative. Oh, you want a bug that does more than talk to you. It would be much better if he made your bed for you every day. Wouldn't it be nice not to ever have to make your bed again? You'd just walk off and leave it—and along would come your special bug and make it for you. Your mother would like that, wouldn't she! A bug who did all her housework for her! Now, that would be a wonderful bug to have—one who chewed up all the germs so nobody would ever get a cold again! I wish you had that bug.

After a few imaginative suggestions have been made, there will be more and more creative ideas. Explain to the class that they will make their bugs out of cut paper.

How big will your bug be? Do you think he will be as tiny as bugs we have on earth? No, probably he will be much bigger. After all, bugs that can do all those great things would have to be bigger than the ones we have here. He can't really be a giant bug, either, because you are going to have to bring him back with you. Move your hands six inches

apart—then a foot apart—then two feet apart. He could be as big as this
—or this—or even this, couldn't he?

Will he look like any bug you have ever seen? Certainly not! This is
a good bug—one that does something so important that you have brought
him all the way from another planet. Make him different from any bug
you have ever seen.

Let groups of children take turns selecting a couple of colors of
paper. Give them scissors and they can begin their bugs immediately. As
you walk about the room, give each child a small piece of newspaper to
do his pasting on, a paste brush, and a bit of paste on a scrap paper;
this will give you an opportunity to see what each child is doing and to
help wherever you are needed.

You may have as many colors as you need for your special bug. If
you need just a little bit, see if someone nearby has some extra that you
may use. If you need a large piece, get it from the supply area. I wonder
what those big things are for. No—no, don't tell me! Wait until your bug
is finished—then you may tell everyone about it. Um-m—blue on blue
doesn't show, does it? Right! Get another color. You don't need to take
those off—just paste the new pieces right on top of them. They look like
big ears of some kind; I wonder if they have some special purpose. I'll
be anxious to hear about what that good bug does. Well, no bug on earth
ever looked like that one, did he? But this isn't an ordinary earth bug.

When all the children have finished their bugs, they will want to
show them; equally important, they will want to tell about them. Comment about some particular part that would help the bug perform his
special job, and compliment a child for an especially creative idea; make
each child feel successful. Later, display all of the bugs.

Some earth bugs may be nuisances, but you won't want to spray any
bug killer on these bugs. Imagine having a bug that could prevent war—
or one to shovel snow for you while you watched television! Bugs can be
good!

Make It Easy—For Yourself!

1. Make sure children have lots of ideas before they begin work.
2. Provide bright colors and black. Eliminate brown and gray to
   encourage children to use nonrealistic colors. Paper cut to 6″ × 9″
   should be large enough and so prevent unnecessary waste.
3. No pencils! No preliminary drawing. Think the shape and then
   cut it.

4. Teach children to do their pasting on newspaper to keep their work and work areas clean.
5. If paste brushes are not available, paste applicators can be made by folding scraps of paper until they are narrow strips, bending them in the middle to give them added strength.
6. As you walk around the room during the lesson, carry a 12″ × 18″ paper as a tray. Let children lay any flat scraps on it that they don't want. This will give them more work space as well as make the final cleanup easier.
7. Have each child put on the back of his bug a brief description of the good thing the bug does. You will want to use it when you display the bug.
8. Encourage older children to look up information in the library about the planets.

lesson 2
Way-Out People

OBJECTIVES

1. To make it desirable for children to let their art differ from that of other children.
2. To encourage children to use found or scrap material in their art.
3. To provide opportunity for working on an extra large project.
4. To permit children to choose an individual or a cooperative project.

You thought you had seen all kinds of people, but you've never seen any way-out people like these!

Have you ever wondered what spacemen look like? Not astronauts who travel through space, but people who live way out in space. They aren't like us at all. If we saw them we would recognize that they were people—but a kind of people we had never seen before. What do you suppose they would look like?

Some children will have immediate ideas, but other children will be hesitant. Draw them into the conversation, too. Must they have two eyes? Would they look like our eyes? How might they be different? Now, that's an interesting idea! The head might be the entire body! Imagine seeing

275

Mr. Space

a person like that! Would they have the same color skin we have? Would all the space people *have* to have green skin? A mouth that big! He must either eat or talk an extra lot!

After many original ideas have been expressed, show the class the boxes you have for them. They look like boxes, but they really aren't— they're parts of space people. Can you put them together to make those people you were talking about? There are some giant boxes, some smaller ones, even some tiny ones. There are long ones and fat ones and even some round ones.

There are other things besides boxes. Have an assortment of paper; paper cups; pipecleaners; large buttons; string-like material such as raffia, yarn, cotton roving. Then, of course, you will need paint to give them whatever kind of skin you desire. Can you begin to see your spaceman?

Explain that they may choose to work alone, or they may choose partners who would like to work with them. If you have a partner, you will probably want to use large boxes to make a giant person. If you decide to work alone, you will want to make a smaller space person. You decide which way you would like to work.

Let the children who want a partner choose someone. Those children who want to work alone can begin to select their materials and go to work immediately. Don't take too many things at one time—you may change your mind about what you need. Give a tube of glue to each child or to each set of partners.

You will be as busy as the children are. First you will want to go to each child to see what plans are being made. There will be questions for you to answer and disputes for you to settle. There will be children who need to be encouraged and given confidence in themselves and what they are doing; you will need to be everywhere at one time.

You must be going to make a large space person with two boxes as big as those! How will he be different from people on this earth? His body will be square, but how else will he be different? That's a good idea. Those paper cups make strange eyes, don't they? How are you going to make him stand up by himself? Even a spaceman would be sick if he couldn't hold himself up! Well, try it and see if it works. Good! Yes, I can see that his mouth moves! That was an excellent idea. Are you ready to paint your spaceman? Wouldn't it be a good idea to paint him before you add the hair? Would you like me to help you hold that?

Have an area for painting supplies. When children are ready to paint, let them choose their supplies and continue work. Remind them to let one color dry before they add another one touching it so the colors won't run together. You can mix colors to make new ones.

Gradually, one spaceman after another will be finished. Find a spot in the room where each one can be left undisturbed until it is thoroughly dry. In the meantime, let any person who worked on a particular spaceman tell about it and show any unusual features. There may be eyes that move on one, a mouth that can be opened and closed on another. A special material may be a part of one spaceman; another may have been given an unusual but appropriate name. Each one will have something about it that is different; be sure you recognize the thought and effort that has gone into each construction.

Where will you display so many large things? Many places! One may serve a special purpose in the cafeteria to display the menu for the week; another might greet visitors in the lobby and point the way to the principal's office; one in the library might announce a special section the librarian plans for books about space. You'll want some in your classroom, of course. Several small ones may fit into a hall display case.

They're different, all right! Nobody ever saw any way-out people like these before! But then—nobody else has ever been that far out in space.

Make It Easy—For Yourself!

1. Encourage children to plan their space people before they begin work so they will have more unusual constructions.
2. Use glue to attach parts; it will be stronger than paste. Glue takes longer to dry than paste, and until then it is weak. Hold the parts together until the glue dries and becomes strong.
3. You may have to take two lessons to complete the project rather than continue work for too long a time.
4. Egg cartons, broken in half, make fine palettes. They can easily be held while the children are painting. Use up to six colors—or save one or more compartments to mix new colors.
5. Have several large cans of water in various parts of the room so one will be easily accessible to every child in the class.
6. Plastic squeeze bottles make fine paint dispensers if your paint does not come packaged that way. Paint left in them remains moist and ready for use from one lesson to another.
7. If possible, have a variety of brush sizes: one-inch enamel brushes for the extra large areas; easel brushes; large wash brushes for the details.

Index

A

Air and wind, 139–150
  blow painting, 141–144
    suggestions, 144
  mural, 147–149
    suggestions, 149–150
Animals, 35–46
  cut paper, 43–46
    suggestions, 45–46
  printing, 37–42
    suggestions, 40–41
Animals of the sea, 153–164
  chalk and charcoal, 155–159
    suggestions, 158–159
  cut paper, 161–164
    suggestions, 163–164
Aquarium, making, 161–164
  suggestions, 163–164

B

Beach, deserted, making, 203–207
Blow painting, 141–144
  suggestions, 144
Boat, construction of, 191–193
  suggestions, 193
Box construction, 191–193
  suggestions, 193
Box painting, 247–251
  suggestions, 250–251
Box space men, 275–277
  suggestions, 277
Bugs, 49–60
  stick and yarn constructions, 55–58
    spider webs, making, 55–58
    suggestions, 58
  wire and pipecleaner, 51–53
    suggestions, 53

Bugs, imaginary, 269–272
  suggestions, 271–272

C

Chalk and charcoal, 155–159
  suggestions, 158–159
Chalk drawings, 135–137
  suggestions, 137
Chalk stencils, 255–259
  suggestions, 258–259
Charcoal, 155–159
  suggestions, 158–159
Circle designs, 241–244
  suggestions, 244
City, 85–96
  newspaper pictures, 87–91
    suggestions, 90–91
  real and abstract, 93–96
    suggestions, 96
Civic center, constructing, 107–110
  suggestions, 110
Clouds, cotton batting, 121–124
  suggestions, 123–124
Collage, 115–118, 213–216
  meaning of, 118, 215, 222
  suggestions, 117–118, 215–216
Compass, use of, 242–243
Cotton batting clouds, 121–124
  suggestions, 123–124
Cotton roving, using to make underwater
  garden, 177–180
Country and suburbia, 99–110
  styrofoam construction, 107–110
    suggestions, 110
  transparencies, 101–105
    suggestions, 104–105
Cut paper animals, 43–46
Cut paper to indicate weather, 129–132
  suggestions, 132

Cut paper to make oceans, rivers and lakes, 185–188
  suggestions, 188
Cut paper for sea animals, 161–164
  suggestions, 163–164

D

Daydreaming, painting ideas of, 29–33
Deserted beach, making, 203–207
Differences in people, studying, 21–22
Doorbell wire, coated, for making three-dimensional bugs, 52
Dry brush painting, 69–73
  suggestions, 72–73

E

Earth, 107–110, 211–222
  animals, 35–46
    cut paper, 43–46
    printing, 37–42
  bugs, 49–60
    stick and yarn constructions, 55–58
    wire and pipecleaner, 51–53
  city, 85–96
    newspaper pictures, 87–91
    real and abstract, 93–96
  collage, 115–118, 213–216
    meaning, 118, 215, 222
    suggestions, 117–118, 215–216
  country and suburbia, 99–110
    styrofoam construction, 107–110
    transparencies, 101–105
  litter collage, 219–222
    suggestions, 222
  people, 19–33
    self-portraits, 21–26 (see also "People")
    tempera painting, 29–33
  plants, 61–73 (see also "Trees and plants")
  rocks and stones, 75–83
    living things, 75–79
    painted, 81–83
  suburbia, 99–110 (see also "Country and suburbia")
  trees and plants, 61–73
    dry brush painting, 69–73
    real and unreal, 63–66
Egg carton as paint palette, 32, 124, 144, 149, 193, 235, 277

F

Fingerpainting, 169–174
  suggestions, 173–174
Fixative, use of hair spray as, 66, 137, 159, 259
Flying things, 113–124
  collage, 115–118, 213–216
    meaning, 118, 215, 222
    suggestions, 117–118, 215–216
  cotton batting clouds, 121–124
    suggestions, 123–124
Fossils, making of with plaster of Paris, 197–201
  suggestions, 200–201

H

Hair spray as fixative, 66, 137, 159, 259

L

Lakes, 183–193 (see also "Oceans, rivers and lakes")
Legend pictures, 261–264
  meaning of, 261
  suggestions, 264
Litter collage, 219–222
  suggestions, 222
Living things, 75–79
  suggestions, 79

M

Mixed media pictures, 203–207
  suggestions, 206–207
Moon, 239–251
  box painting, 247–251
    suggestions, 250–251
  circle designs, 241–244
    suggestions, 244
Mural, 147–149
  suggestions, 149–150

N

Negative stencil, meaning of, 256
"New" animals, making, 43–46
Newspaper pictures, 87–91
  suggestions, 90–91

O

Oceans, rivers and lakes, 183–193
  box construction, 191–193
    suggestions, 193
  cut paper, 185–188
    suggestions, 188

P

Painted rocks, 81–83
  suggestions, 83
Pencils, avoiding use of, 26, 32, 40, 66, 79,
    83, 96, 117, 132, 137, 158, 188, 215,
    236, 250, 259, 264, 271
People, 19–33
  self-portraits, 21–26
    differences in people, studying, 21–22
    making, 21–26
    suggestions, 26
  tempera painting, 29–33
    suggestions, 32–33
Pipecleaner bugs, 51–53
Planets, other, and space, 267–277
  box space men, 275–277
    suggestions, 277
  bugs, imaginary, 269–272
    suggestions, 271–272
Plants, 61–73, 167–180
  dry brush painting, 69–73
    suggestions, 72–73
  fingerpainting, 169–174
    suggestions, 173–174
  stitchery, 177–180
    suggestions, 180
Plaster of Paris, using to make fossils, 197–
    201
  suggestions, 200–201
Plastic squeeze bottles as excellent paste
    dispensers, 235–236, 277
Polymer, painting with, 203–207, 247–251
Positive stencil, meaning of, 256
Pretending to be someone, painting idea
    of, 29–33
Printing, 37–41
Protective coloration, 77

R

Rivers, 183–193 (see also "Oceans, rivers
    and lakes")

Rocks and stones, 75–83
  living things, 75–79
    suggestions, 79
  painted, 81–83
    suggestions, 83
Rouault, Georges, 21–22
Roving, cotton, using to make underwater
    garden, 177–180

S

Sand, shells and, 195–207 (see also "Shells
    and sand")
Sea, animals of, 153–164
  chalk and charcoal, 155–159
    suggestions, 158–159
  cut paper, 161–164
    suggestions, 163–164
Self-portrait, making, 21–26
Shadows, 227–230
  suggestions, 229–230
Shapes, changing, 227–230
  suggestions, 229–230
Shells and sand, 195–207
  mixed media, 203–207
    suggestions, 206–207
  plaster of Paris, 197–201
    suggestions, 200–201
Silhouettes, 101–105
  suggestions, 104–105
Sky, 111–150
  air and wind, 139–150
    blow painting, 141–144
    mural, 147–149
  flying things, 113–124
    collage, 115–118, 213–216
    cotton batting clouds, 121–124
  weather, 127–137
    chalk drawings, 135–137
    cut paper, 129–132
  wind, 139–150 (see also "Air and
    wind")
Skylines, 87–91
Space, 267–277
  box space men, 275–277
    suggestions, 277
  bugs, imaginary, 269–272
    suggestions, 271–272
"Spider webs," making, 55–58
Squeeze bottles, plastic, as excellent paint
    dispensers, 235
Stars, 253–264
  chalk stencils, 255–259
    suggestions, 258–259
  legend pictures, 261–264
    meaning of, 261
    suggestions, 264

Stencil, meaning of, 256
 making, 255–259
 suggestions, 258–259
Stick constructions, 55–58
Stitchery, 177–180
 suggestions, 180
Stones, 75–83
 living things, 75–79
 suggestions, 79
 painted, 81–83
 suggestions, 81–83
Storm, depicting, 135–137
 suggestions, 137
Straws, use of in blow painting, 141–144
Styrofoam construction, 107–110
 suggestions, 110
Suburbia, 99–110
 styrofoam construction, 107–110
 suggestions, 110
 transparencies, 101–105
 suggestions, 104–105
Sun, 225–236
 changing shapes, 227–230
 suggestions, 229–230
 tempera painting, 233–236
 suggestions, 235–236

T

Tempera painting, 29–33, 233–236
 suggestions, 32–33, 235–236
Tongue depressor, use of in blow painting, 141–144
Transparencies, 101–105
 suggestions, 104–105
Trees and plants, 61–73
 dry brush painting, 69–73
 suggestions, 72–73
 real and unreal, 63–66
 suggestions, 66

U

Underwater garden, making, 177–180
 suggestions, 180
Universe, 209–277
 earth, 211–222
 collage, 115–118, 213–216
 litter collage, 219–222
 moon, 239–251
 box painting, 247–251
 circle designs, 241–244
 planets, other, and space, 267–277
 box space men, 275–277
 bugs, imaginary, 269–272

Universe (*cont.*)
 stars, 253–264
 chalk stencil, 255–259
 legend pictures, 261–264
 sun, 225–236
 changing shapes, 227–230
 tempera painting, 233–236

V

Varnishing rocks, 81–83

W

Water, 151–207
 animals of the sea, 153–164
 chalk and charcoal, 155–159
 cut paper, 161–164
 oceans, rivers and lakes, 183–193
 box construction, 191–193
 cut paper, 185–188
 plants, 167–180
 fingerpainting, 169–174
 stitchery, 177–180
 shells and sand, 195–207
 mixed media, 203–207
 plaster of Paris, 197–201
Water plants, painting, 169–174
 suggestions, 173–174
Wax crayons, use of to make transparencies, 104
Weather, 127–137
 chalk drawings, 135–137
 suggestions, 137
 cut paper, 129–132
 suggestions, 132
Wind, 139–150
 blow painting, 141–144
 suggestions, 144
 mural, 147–149
 suggestions, 149–150
Wire bugs, making, 51–53

Y

Yarn, using to make underwater garden, 177–180
 suggestions, 180
Yarn constructions, 55–58